To Daddy

with ... m

Elizabeth

Josephine and Esther

C. C. DOWNS

ex-RAF

ex-MNT Co Ltd

Radar Section

Radar Days

Frontispiece Sir Henry Tizard, the genius who foresaw how the Battle of Britain would be fought and made sure that the tools for victory were provided.

Radar Days

by

E G Bowen CBE, FRS

Foreword by
Lord Bowden of Chesterfield

Adam Hilger, Bristol

British Library Cataloguing in Publication Data

Bowen, E.G.
Radar days
1. Airplanes—Radar equipment—History
I. Title
629.132′51 TL696.R2

ISBN 0-85274-590-7

Consultant Editor: **Professor A J Meadows,**
Loughborough University

Published under the Adam Hilger imprint by IOP Publishing Ltd
Techno House, Redcliffe Way, Bristol BS1 6NX, England

Printed in Great Britain by J W Arrowsmith Ltd, Bristol

Contents

Foreword

Few people today realise what terrible dangers threatened this country about half a century ago when Hitler had begun to rearm Germany and build up an enormous air force. Stanley Baldwin, who was Prime Minister at the time, remarked that 'The bomber will always get through.' He gave the impression that we should try to appease Hitler, but other people, most notably Winston Churchill, were determined to resist Hitler by all possible means. In this book Dr E G Bowen describes how airborne radar helped to defeat Hitler and the bombers of the Luftwaffe. It was a very close contest.

In December 1934 Sir Henry Tizard, Rector of Imperial College, London, set up a small committee for the scientific survey of air defence and recruited some of the most able scientists in the country to help him consider ways of detecting approaching aircraft. The Tizard Committee consulted Mr R Watson Watt who was the Director of the Radio Research Station in Slough, one of the best equipped and certainly one of the most well known establishments in the world. Watson Watt told Tizard that an aircraft must reflect energy from any powerful transmitter on the ground. He organised an experimental test of this prediction by arranging for an aircraft to fly near the BBC's transmitter at Daventry, and, sure enough, the receiver on the ground detected echoes from the aircraft. This demonstration convinced Sir Hugh Dowding, future Commander-in-Chief of Fighter Command, that the idea had merit. He persuaded the Treasury to pay for the development of a system to make use of these echoes and to help Watson Watt recruit more staff. Dr Bowen was the first man to join Watson Watt.

Watson Watt and his colleagues showed how it was possible to detect formations of enemy aircraft hundreds of miles away and to estimate their height and direction and the number of aircraft in each formation. It was then decided to coordinate the observations from all the stations which were observing enemy formations. The RAF built several 'filter rooms' where the results could be plotted and from where, it was hoped, Fighter Command could control our own squadrons.

While these filter rooms were being built, a barrage balloon dragged its moorings and drifted over London. Sir Hugh Dowding discovered to his horror that it took Fighter Command half a day to find the balloon and shoot it down. Dowding reacted very swiftly, as he always did, so that when the Battle of Britain began the control and direction of Fighter Command had been transformed. The Germans were amazed. They had a radar system of their own and their individual sets were quite good, but they had no organisation to make adequate use of them. They had never imagined that a system like ours could be built. In fact they confidently expected that they had enough aircraft to overwhelm the British defences, in particular Fighter Command, before they invaded. Churchill, who had become Prime Minister by this time, remarked after the battle that 'Never in the field of human conflict was so much owed by so many to so few!' Everyone in the country knew that Churchill was right, but very few people knew at the time that, had it not been for our radar system, the 'few' would almost certainly have been defeated.

Even before the Battle of Britain had been won, Sir Henry Tizard decided that while we could defeat the Germans in daytime, we had no defence against bombers which came by night. He urged Bowen to build airborne radar sets which could detect incoming night bombers. Bowen accepted the challenge and devoted himself to the development of airborne radar for the rest of the war. This work was astonishingly successful and had a decisive effect on the war in the air and at sea. He quotes, with very great satisfaction, from the report written by the commander of a German night fighter squadron who complained that British night fighters could pick off his men at will 'like currants in a cake'.

The same was true in the war against German submarines. They were sinking so many of our merchant ships that we might have been starved into surrender, unless they could be sunk in large numbers. This, too, was achieved by airborne radar after a long-drawn-out battle. Admiral Doenitz, Commander-in-Chief of the German Navy, said that those who created airborne radar were the 'saviours of their country'. No one was in a better position than Doenitz to appreciate what Bowen and his colleagues had done.

As soon as the idea of radar had been proposed by Watson Watt, he realised the enormous importance of deciding whether an aircraft was a friend or foe. The system which was produced was called IFF (Identification of Friend from Foe), but the problems of identification were never completely solved. The failure of the first system was demonstrated in a most spectacular way on the very first day of the war. While the Prime Minister was speaking, just as he announced that we were at war with Germany, a stray French aircraft came over the Channel. It was detected and assumed to be hostile—the sirens went and everyone had to go into air-raid shelters. To overcome these problems of identification a series of IFF sets were introduced. The first sets operated by detecting the signal received from the ground station and then re-radiating it in a modified form.

The first set to go into mass production was the so-called Mark II. It had to cope with transmitters in the main chain, as well as other transmitters on half a dozen other frequencies (such as gun-laying sets) and portable transmitters which had been designed to replace the main chain if it was ever destroyed by enemy action. The sets themselves were very complicated and the system was never entirely satisfactory. It was replaced by a new and much more versatile system called IFF Mark III. I was made responsible for the development and production of these sets. Shortly after, the Commander-in-Chief of Fighter Command sent for me and said: 'You are in charge of this new system of identification, aren't you?' 'Yes, sir', I replied. 'Well, yesterday when a Stirling bomber was returning from a raid on the Ruhr it was thought to be hostile and so a couple of Beaufighters were sent to intercept it. The first Beaufighter shot the Stirling down, and then was itself shot down by the other Beaufighter. Two aircraft and their crews were lost because this system of yours didn't work. What are you going to do about it?'

Many years later, when we were both members of the House of Lords, I reminded him of this conversation. He remembered it well: 'I thought it was a bit unfair at the time, but it certainly got you activated in a hurry!' It most certainly did. After we had 'tamed' the sets we staged a demonstration of the system, to which we invited American observers. The system worked very well; the Americans decided to adopt it and put the sets into mass production.

As background information for the readers, I wish to record here some personal reminiscences about the protagonists and what happened to them after the war.

When the war was over A P Rowe, who was the Chief Superintendent of the Telecommunications Research Laboratory (TRE), decided to go to Australia for the sake of his health and he became Vice-Chancellor of the University of Adelaide. To his dismay he discovered that there was no proper pension scheme in the University and many elderly professors went on working long after they should have retired. He persuaded the State Government to introduce a proper system of pensions and to make it possible for staff to retire at the age of 65. This was very beneficial, and the University was very grateful. But then he tried to run the University just as he had run TRE. The Senate rebelled, and Rowe returned to Malvern where he taught in the College which had housed TRE during the war. He used to boast that if the Battle of Waterloo had been won on the playing fields of Eton, the Battle of Britain had been won on the playing fields of Malvern. He mellowed very much as he got older. For example, he used to tell visitors that he once sent for me to go to his office in a hurry. I had to run uphill and then climb several flights of stairs. I was completely winded when I arrived. 'Tell me, Bowden, what do you do when you feel you need some exercise?' he said. 'Oh, I lie down until the feeling passes!'

The story of the last days of Watson Watt is so tragic that I must recount it, if only briefly. At the end of the war he seemed to be riding high and it seemed obvious that he would be much involved in the formulation of plans for the control and navigation of aircraft all over the world. He applied to the Royal Commission for Awards to Inventors for an award for the work he had done and the inventions he had made during the war. In the course of his evidence he explained to the Commissioner, Lord Justice Cohen, that he had been the first person to have used women in the filter rooms. 'Are you claiming, Sir Robert,' said the Judge 'to have invented women?' Be that as it may, he and those of his colleagues who joined with him in the application were awarded the sum of £101 000, with Watson Watt, Wilkins and Bowen as the principal recipients. However, shortly after this he made a sudden change of plan; he was divorced by his wife and left Britain for a new career in Canada. Unfortunately, he had great difficulty adjusting to civilian life. During the war he had been imaginative and creative and almost ruthless in getting his own way. These were the qualities which had made him great during the war, and which made radar possible. They appeared to have deserted him after the war and he never settled down to a peacetime career. After several changes of fortune, he finally died in an old people's home in Scotland, unknown and apparently forgotten.

I must add a few words about A F Wilkins, who had worked closely with Watson Watt at the Radio Research Station in Slough. Wilkins was primarily responsible for constructing the Home Chain, which helped defend the Thames estuary during the Battle of Britain. He extended the Chain all over the country. It was he who chose the sites, took the responsibility for buying them, and made sure that proper access roads were built. All this was difficult enough, especially in wartime, but he was firmly instructed by the Air Ministry not to build radar stations where they might interfere with grouse shooting! By the time the Chain was finished, TRE had recruited a great number of graduates from universities. They often ignored Wilkins, but he never seemed to mind. He knew what he had done and, in a modest way, he was rather proud of himself. I am very glad to have been able to count him among my friends.

I recall also that Sir Hugh Dowding, who achieved so much in winning the Battle of Britain, was dismissed almost as soon as the battle was over. He received a miserable pension and died in poverty. Sir Henry Tizard, whose Committee first inspired the development of radar, both airborne and ground-based, and who led the Tizard Mission to America, was dismissed as soon as he returned to this country. To add to these seemingly ungrateful actions, after the war Churchill's government was defeated in a general election. He spoke for many when he remarked very soberly that: 'Ingratitude to their Great Men seems to be common in democratic societies.'

Lord Bowden of Chesterfield

Preface

It is now more than 50 years since radar began in Britain. A technical account of the subject was given in a collected work by the Institution of Electrical Engineers and in several textbooks which appeared soon after the war. These were followed by a few personal accounts of the radar years, like those by Watson Watt and A P Rowe. Then came a gap of nearly 40 years during which many parts of the story remained untold. Airborne radar has become one of the most important branches of civilian and military radar, but the way in which it began has never been described in a comprehensive fashion. In much the same way, the all-important Tizard Mission to America in 1940 has only been described in a fragmentary manner. It happens that I was involved in both airborne radar and the Tizard Mission and the main purpose of the present book is to fill in the missing parts of the story.

The early accounts of British radar were usually written from the viewpoint of someone in the higher echelons of government or as seen by the management. The present book is from the point of view of the individuals who did the work at the laboratory bench followed by flight trials in which the lone experimenter tried to simulate, as best he could, the problems likely to be met in the battles yet to come. This was a very demanding activity which seldom took place under the safest or the most comfortable conditions. It was not a pursuit coveted by everyone but, as we shall see, it had special rewards for those who could survive the pressures.

In writing so long after the event, it is inevitable that hindsight will come into an account of this kind. I have done my best to concentrate on the story as it appeared to us at the time, with all its successes and failures. I should also emphasise that this book does not pretend to be a complete history of radar; it is strictly a story as seen by one person following a very special track—that of airborne radar. It is hoped that it will encourage others to give their own account of events as they saw them; only in this way will the complete story finally be told. Throughout the book I have referred to the new science as *radar*—a term which is universally understood. This terminology did not come into use until quite late in the war, but to avoid confusion I have used the word *radar* as if it were employed from the outset.

Acknowledgments

In writing this book, I have received an enormous amount of help from many sources. First of all, I want to thank the members of my original airborne team at Bawdsey Research Station for their friendship over the years and for reminding me of many anecdotes about the early days. I particularly want to thank Keith Wood for being so helpful and for making available his diaries which are our only record of the early radar flights. Similarly, I am grateful to Hanbury Brown for access to his records and for a great deal of assistance with the manuscript. I also want to record my appreciation for much help received from Air Commodore H A Probert of the RAF Air Historical Branch and Guy Hartcup, who was once on the staff of that branch. Without the ready assistance of all these individuals, this work could never have been written.

I cannot praise too highly the staff of the Public Records Office at Kew for their help and perseverence in producing old radar documents. For those who are not aware of the fact, that Office is a copious and rewarding source of documents about early radar history. I also want to express my appreciation to Miss Sally Atkinson for the meticulous way she has preserved my old papers and photographs, and for an enormous amount of help in the final preparation of this book. My special thanks are due to Miss Patricia King of Washington DC for her remarkable facility in producing, as if by magic, documents dealing with the early days of the Tizard Mission. Similarly, I am indebted to Tony Brinkley of the Australian National University in Canberra for checking my memories of Orford and Orfordness and for a most thorough and painstaking reading of the manuscript. In addition, I wish to acknowledge, with many thanks, Sir William Hawthorne's kindness for the award of a Fellowship at Churchill College which allowed me to study old radar documents, and a grant from the Royal Society which made it possible for me to travel to Britain for this purpose in 1979. Finally, I am grateful to Beth Nanlohy and John Masterson for creating the cover design.

E G Bowen

1

The Beginning of Radar in Britain

It all began in the early spring of 1935 when Mr R A Watson Watt, the Superintendent of the Radio Research Station at Slough, made a suggestion which was destined to have a decisive effect on the hostilities which began four years later and reached a climactic stage during the Battle of Britain in 1940. The proposal he put forward was that warning of the approach of enemy aircraft should be obtained by sending radio waves in the direction of the oncoming aircraft and that knowledge of their approach should then be obtained by listening-out for the minute signals reflected back from the metal structure of the aircraft. The suggestion was a bold one as the energies involved were extraordinarily small. The energy returned from an aircraft in this way would be less than a million, million, millionth part of that sent out—in engineering terms the efficiency would be literally 10^{-19}. There were any number of people who could prove by arithmetic that this could not be done and that it would be foolish even to try.

To understand the importance of Watson Watt's suggestion it is necessary to go back a few years to the time when Adolf Hitler came to power in Germany. Under his guidance, the Nazis took a firm grip of the political and military life of the nation. Germany's military strength built up at an unprecedented rate and a formidable air force was in the making. No attempt was made to conceal these developments. On the contrary, at numerous rallies, Hitler was at pains to publicise his objectives and make his intentions crystal clear. With ringing oratory and mesmeric force, he whipped his followers into a frenzy which could only have one outcome—one in which the Nazi creed would spread like a plague over the rest of Europe.

In Britain there was a very mixed reaction to these events. A few people were thoroughly alarmed, but the great majority were unmoved and apathetic—they preferred to put the awful thought out of their minds and hoped that the problem would go away. After all, had not the prestigious Oxford University Debating Society carried a motion in February 1933 that 'This house will under no circumstances fight for King and Country'? There were even people in high places who seemed to approve of what was going on—the notorious 'Cliveden Set'—the Astors, the Moseleys and the Mitfords, who hob-nobbed with leading Nazis and lionised them. Even the London 'Times'—the Thunderer—seemed to give tacit approval to the revolutionary changes taking place in Germany. In 1935 the playboy Prince of Wales suggested to the British Legion that they should 'stretch forth the hand of friendship' to Nazi Germany and in December 1936 when he abdicated his throne, Unity Mitford was heard to say 'Hitler will be terribly unhappy about this. He wanted Edward to stay king'.

Nevertheless, the country was not without people who were deeply concerned about the problem and among these were two people who were destined to play a crucial part in the coming struggle. The first of these was Winston Churchill. At that time he was in the political wilderness, a peripatetic, restless figure, out of touch with his contemporaries and to a large extent out of touch with the man in the street. Even so, he knew with certainty what the rising tide of Nazism meant and he directed all his thoughts and energies to countering it.

The other was Henry Tizard—a man uniquely qualified to judge the situation for himself and to map out the course of action he should take. He was a chemist by training with a bias towards thermo-dynamics and in this capacity he had spent a year working in Nernst's laboratory in Berlin in 1908. He knew the German people and had first-hand experience of that country preparing to go into a war. During World War I Tizard served in the Royal Flying Corps and in 1917 was in charge of the scientific work of the Aeroplane and Armaments Experimental Station, first at Orfordness and then at Martlesham Heath in Suffolk. This was a unique establishment devoted to testing the performance of aircract and the weapons they carried in a scientific way.

Between the wars Tizard held a variety of high governmental posts, perhaps the most important of which was as Secretary of the Department of Scientific and Industrial Research. In this position he was in close touch with the principal academic and governmental research establishments of the day; he knew the people involved in an intimate way and the strengths and weaknesses of the researches being undertaken. He was deeply conscious of the interplay between scientific research and government policy and well aware of the importance of deploying men and money to plug an important gap or to open up a new field of endeavour, whether for economic benefit or for the defence of the country. Lastly, Tizard was very much aware of the length of time which must elapse between initiating something at the research level and its practical application to a specific problem. In those fields in which he was destined to play such an important part he knew that it was useless to wait for disaster to strike. He soberly and scientifically worked out the implications well ahead of time and signalled which way to go. Alternatively, if he himself could not point the way, he knew where to turn to find the answer.

In 1934, therefore, Tizard was in a unique position to consider the problem of Great Britain's air defence. At that time he was the Rector of Imperial College, a position which gave him leisure to think about the problems England would soon have to face and ready access to governmental and military circles in London. Towards the end of 1934 Tizard began to thank deeply about the problems of air defence and what could be done about it. He realised that in a few years' time Britain would be subjected to a devastating air attack and, as things stood, the country was defenceless against it. Tizard disagreed fundamentally with Stanley Baldwin's dictum of 1932 that '. . . the bomber will always get through' and was convinced that both aircraft and anti-aircraft guns could be made effective against such an attack, but only if they were in the right place at the right time. In other words, warning of approach of enemy aircraft was a critical requirement, coupled with a precise indication of where the enemy aircraft were at any particular moment and where they were heading.

A committee was formed to consider this problem. It consisted of P M S Blackett, A V Hill and H E Wimperis, with Tizard as Chairman and A P Rowe as Secretary. Blackett was a Professor of Physics at Birkbeck College who had served as an engineering officer in the Navy in World War I. He became a valuable members of Rutherford's nuclear team at Cambridge and had already done the work which was to earn him a Nobel Prize after the war. Hill was a distinguished physiologist, also from Cambridge. who already held the Nobel Prize. In World War I he had been an artillery officer with extensive experience of gun ranging using acoustic methods. The prior experience of Hill and Blackett in the Army and the Navy nicely complemented that of Tizard in the field of aeronautics. Wimperis was a personal friend of Tizard's and was the first Director of Scientific Research at the Air Ministry, while Rowe was his personal assistant.

The committee was formed at the instigation of Wimperis, but I have a hunch that it was Tizard himself who first suggested it. It was called the Committee for the Scientific Survey of Air Defence and was formally responsible to the Committee for Imperial Defence. It

Figure 1.1 Sir Robert Watson Watt, the father of British radar. He was the man who proposed the chain air-warning system and saw it through to completion.

became known as the Tizard Committee and the first meeting was held in January 1935. There were, at that time, any number of suggestions about air defence ranging from those which were barely possible to others which were slightly eccentric to still more which were totally irrational or else outright fraudulent. People like Professor Lindemann of Oxford University, for example, advocated wires trailing from aircraft or drifting balloons. The kindest thing one can say about most of these schemes is that in spite of much discussion and many expensive trials, they never led to a workable method of defence.

One idea which appeared to have a special fascination for the Air Ministry was the concept of a death ray which would either claw an aircraft out of the sky or burn up the occupants at the turn of a switch. These proposals were never based on recognised physical principles, nor were they described in sufficient detail for a proper assessment to

Figure 1.2 Arnold F Wilkins who first calculated the intensity of a radio wave which would be reflected from an aircraft and conducted the Daventry experiment which demonstrated the feasibility of radar. He was the leader of the exceedingly small group which pioneered British radar at Orfordness and Bawdsey in 1935.

be made. Their proponents usually referred to black boxes which generated weird mixtures of ultraviolet rays, x-rays and radio waves, the combined effect of which was supposed to produce dire results. The schemes had two things in common: one was the claim that sheep or pigeons had dropped dead on being subjected to the mysterious rays; the other was that vast sums of money would be required before the inventors could proceed with their death-dealing devices.

One more such project was submitted to the Air Ministry about this time and it was sent to Watson Watt at Slough for an assessment. The subsequent story has been told many times—how he consulted one of his staff, Arnold Wilkins, and asked him to calculate how much radio energy would be required to damage an aircraft or adversely affect the crew. The answer came out at an impossibly high figure. However, in the course of making the calculation, Wilkins realised that if radio waves were sent in the direction of an approaching aircraft, while no damage would be done, there was a possibility of doing something quite different, namely detecting the radio waves reflected back from that aircraft.

This proposal was put forward in a famous memorandum which Watson Watt submitted to the Air Ministry on 12 February 1935. It fell on very receptive ground. Tizard was already saying that the essence of the air defence problem was to obtain warning of approach of enemy aircraft. Now came Watson Watt with the proposal that this warning could be obtained by detecting the radio waves reflected from approaching aircraft. This was the trigger which began a whole sequence of events which were to have a profound effect on the conduct of war, and which continue to do so to the present day.

Not that the idea of reflection of radio waves from an aircraft was original in itself. Englund, Crawford and Mumford, three engineers of the Bell Telephone Company in the United States, had noted that signals reflected from an aircraft passing overhead could be detected by a sensitive radio receiver and they described this in a paper in the *Proceedings of the Institute of Radio Engineers* in 1933. A year earlier, some British Post Office engineers—curiously enough one of them was also called Mumford—made a similar observation and reported it in a *Post Office Report* which did not receive wide circulation. We now know that Pat Hyland of the US Naval Research Laboratory made a precisely similar observation even earlier. This was at Bolling Field in Washington in 1930, as a result of which he proposed that both ships and aircraft could be detected using the reflection of radio waves. This proposal did not find favour with the Navy Department and no money was allocated to the project. Hyland felt so badly about this that he resigned from the Naval Service a few years later.

It is worth noting that even in the late 1930s there were any number of countries which did not feel that they were threatened by air attack and, therefore, did not act on the observations freely published in 1933. Britain, on the other hand, faced a serious threat and was forced to take action. It was the Tizard Committee which defined the problem in a clear way and expressed the need for early warning of the approach of enemy aircraft and it was Watson Watt who suggested the solution. Not only did the Watson Watt proposal specify how to provide warning, it promised precise location as well. What was even more important, the merit of the proposal was recognised by the Government in 1935 and acted upon.

At this time, Watson Watt was a scientist with twenty years of experience of government research. He was born at Brechin in Aberdeenshire in 1892 and took an initial degree in Physics and Engineering at Dundee a few years before the outbreak of World War I. After offering his services to the Army in 1914 without response, he joined the Meteorological Office in 1915. He began work on the detection and location of thunderstorms at the Royal Aircraft Establishment in Farnborough and remained there throughout the war. The aircraft of the day were extremely frail and even flying in cumulus cloud was considered hazardous. Active thunderstorms, especially, were to be avoided and there was a pressing need to locate and map dangerous thunderstorm regions. Thunderstorms were known to be a prolific source of atmospherics and Watson Watt began the design of an elementary radio direction-finder which gave the general position of such storms. Research continued after the war and reached a high point in 1922 when sealed-off cathode-ray tubes first became available. With assistance from J F Herd, a fellow Scot who had joined him at Farnborough, he constructed an instantaneous direction-finder using the new cathode-ray tubes which gave the direction of thunderstorm activity. It was admittedly of low sensitivity, but it worked and was the first of its kind.

He did much of this work at a new site at Ditton Park, near Slough, where the National Physical Laboratory (NPL) also had a field station devoted to radio research. In 1927 the two endeavours were combined as the Radio Research Station, Slough; it came under the general supervision of the NPL, with Watson Watt as the first Superintendent. This became a centre with unrivalled expertise in direction finding using the cathode-ray tube and in studying the ionosphere using radio waves. A comprehensive account of the technical work of that station is given in a book entitled *The Cathode Ray Tube in Radio Research* by Watson Watt, Herd and Bainbridge-Bell (1933). It is required reading for anyone interested in the prehistory of radar research. Among other things, it shows how the Plan Position Indicator—the PPI of later radars—was very nearly, but not quite, invented in 1933.

I first met Watson Watt in 1933 while I was a student at King's College in London. He had the reputation of being a very relaxed and pleasant administrator who kept closely in touch with the technical work of the Station. He was much loved by the staff and with his deputy, Jimmy Herd, headed an extraordinarily powerful group of people who were at the forefront of radio research.

Through his membership of the Radio Research Board, Watson Watt was well connected with the Chief Signals Officers of the Army, Navy and Air Force and was frequently in touch with academics like

Professor Appleton of King's College. He had a running battle with the management of NPL, Teddington, as to whether the prime function of the laboratory was fundamental research or the practical application of radio technology. At that time, there was a keen governmental interest in connecting up the far-flung Empire by air and Watson Watt was anxious to get involved in air navigation in support of this. He did not get a good response from the management at Teddington and this had an important consequence in 1935. When Watson Watt made his proposals for detection of aircraft, the NPL were simply not interested in following it through in a practical way and were glad enough to pass responsibility to the Air Ministry.

By a happy coincidence, when Watson Watt first advanced his proposal the Air Member for Research and Development at the Air Ministry was the great Air Vice-Marshal Sir Hugh Dowding, the man who was destined to play a vital role as Commander-in-Chief of Fighter Command during the Battle of Britain. The decision whether to provide money for the new proposal rested squarely on his shoulders. In his gruff way, his first reaction to Watson Watt's memorandum was that he was not much impressed by calculations, but if a practical demonstration were given he might be convinced. This led to the historic occasion on 26 February 1935 when Watson Watt asked A F Wilkins to repeat the Post Office observation of reflection from a nearby aircraft. In an historic experiment using the Daventry transmitter and a Heyford bomber as target, Wilkins did this with clear and irrefutable results. Wheels started to turn and a sum of £10 000 was granted for initial work. A small staff was drawn from the Radio Research Station at Slough and sworn to utmost secrecy. It was at this point that I first came into the picture as the most junior of the three scientists assigned to the task—Wilkins, Bainbridge-Bell and Bowen.

Born in Wales, I had taken an honours degree in Physics at University College in my home town of Swansea. From the age of eleven or twelve years, when broadcasting first began in Britain, I had taken an obsessive interest in radio. Professor E J Evans, my kindly professor at Swansea, recognised this and arranged for me to take a PhD under Professor Appleton, then the Professor of Physics at King's College in London and the doyen of radio research in Britain. As part of my research, Appleton arranged for me to spend much of my time during 1933 and 1934 using the cathode-ray direction-finder at Slough. In this way I became familiar with the work of the Station and on friendly terms with the staff. What is probably more important, I came under the benign influence of Watson Watt.

When Watson Watt's proposal was accepted by the Air Ministry it was obvious that more staff would be required and the usual advertisements were issued. I was told about this and advised to apply. Britain was still feeling the effects of the Depression and such jobs were hard to come by. There was a long list of applicants and a shortlist of about eight, of which I was one. I remember a very relaxed interview with Watson Watt and his deputy, Jimmy Herd, during which the main item discussed was whether, as a Welshman, I could survive with two Scotsmen as bosses. They challenged me to sing the Welsh national anthem as the price for entry into the system. I responded by saying that I would sing the Welsh national anthem if they would first sing the Scottish anthem. I did not know for sure, but I guessed it could only be done with the aid of bagpipes. Needless to say, they did not take up the challenge and I do not know what the Scottish anthem sounds like to this day.

Mine was the only appointment made and I joined the staff of the Radio Research Station towards the end of April 1935 as a Junior Scientific Officer at the princely salary of £256 per annum. I was still unaware of the fact that the Radio Research Station harboured a great secret, but I realised something very unusual was in the wind when, on my very first day, I was introduced to the provisions of the Official Secrets Act. This task was solemnly performed by Jimmy Herd and I well remember the relish with which he explained the penalty for the slightest deviation from the most meticulous standards of security—literally to be 'hanged by the neck until life was extinct'. Jimmy Herd enlarged on this with enormous gusto and only after I had agreed to the provisions and signed on the dotted line was I told the secret of radar. Needless to say, I was enormously impressed by the demands of my new job.

I was given a copy of Watson Watt's memorandum to read overnight. It was in double envelopes and I had strict instructions not to let it out of my sight. I was staying for a few days at the Manor Hotel at Datchett. That night I ate a solitary dinner with the envelope concealed in an evening newspaper on the table in front of me. After dinner, I went back to my room to study the manuscript. Before retiring to bed a few hours later, I felt the need for a nightcap and realising that I could not take the envelope down to the bar, I looked around for a good place to conceal it. Clearly, a drawer or a cupboard was not good enough. I seized on what I thought was a simple solution and put the envelope between the sheets at the bottom of the bed. It was quite flat and could not possibly be noticed.

I came back to my room, prepared for bed and checked that the memorandum was safely in its place; sure enough it was still in the same position, with a hot-water bottle placed neatly on top of it! That night I had nightmares about swinging on the end of a gibbet on some lonely heath outside London. Next morning I made a point of chatting with the chambermaid and was relieved to find she bore no resemblance to a Mata Hari but was a buxom country girl with an unmistakable Buckinghamshire accent.

Watson Watt's memorandum spelled out the fact that a practical scheme of air detection required a measurement of the range of an approaching aircraft. A precise method of doing this was already available from previous work on the ionosphere. The ionosphere is a reflecting layer about sixty miles up in the upper atmosphere which, under suitable conditions, acts as a perfect mirror for radio waves. This was a field in which the Radio Research Station at Slough had specialised for many years and they had all the technology. Briefly, the technique was to send short pulses of radio energy in a vertical direction and to measure the time taken for them to travel up to the ionosphere and return to earth. The elapsed time gave the height of the ionosphere with considerable accuracy. This was a method invented by Breit and Tuve in America in 1925 and was a well established research procedure in many countries around the world.

As already mentioned, doing the same thing with a signal reflected from an aircraft would be much more difficult. The reflected signal would be only 10^{-19} the strength of that sent out and many people considered this to be impossible. A little consideration showed that there were three directions in which to go for improvement. Firstly, a massive increase of something like a hundred times in transmitted power would certainly be required; secondly, as much additional sensitivity as possible should be squeezed out of the receiver; lastly, as much gain as possible should be obtained from the antennae.

The ionospheric transmitters of that era had a peak power output of about 1 kilowatt with a pulsewidth of 200 microseconds. The requirement for aircraft detection was an increase of 100 times in power, with a corresponding reduction in pulsewidth to about 20 microseconds. The more conservative electrical engineers of the day doubted whether any such increase in power was possible but this did not deter the pioneers and with characteristic verve we set out with 100 kilowatts as the target. The receivers of that day were already about as sensitive as they could be made so that an improvement by a factor of two would represent a major step.

The third area in which the signal might be augmented was to exploit antenna gain, but at the long wavelengths contemplated—50 metres in the first instance—little could be expected for a number of years. Virtually the only thing to do was to increase antenna height, which would concentrate the signal towards the horizon instead of throwing it in an upward direction.

It was along these lines that the pioneers focused their attention and the rough outline of an experimental set emerged. In order to preserve secrecy, it was decided to erect the first system not at Slough but at a remote site at Orfordness about 90 miles north-east of London. This was a long spit of salt marsh and shingle extending 15 miles down the coast from Aldeburgh and reached by boat from the village of Orford. It was already owned by the Air Ministry and had been a busy establishment during World War I. However, in 1935 it was run down and virtually abandoned. Some World War I huts were hurriedly re-roofed and reconditioned and became the transmitting and receiving huts respectively. Power was laid on from the mainland and other preparations were made to accommodate the incoming group from Slough.

Orfordness

The radar group left Slough for Orfordness on a delightful spring morning, 13 May 1935. The party consisted of only four people, Wilkins, Bainbridge-Bell and myself of the scientific staff, and George Willis who had been Bainbridge-Bell's technical assistant for many years. Wilkins and I went in Wilkins's Armstrong Siddeley, while Bainbridge-Bell and Willis went in B-B's newly acquired MG. The convoy was completed by two large RAF vehicles loaded with our equipment and we arrived at Orford that evening without incident. Early next morning we gathered at the Orford quay to start transporting the gear across to the island. We were greeted by hail and sleet and a howling east-coast gale. After the mild spring weather we had experienced at Slough, the contrast was alarming and it was some time before the impression that we had arrived at an Arctic waste faded away.

We spent the next few days like slaves, humping the equipment off the wharf at Orford, on to a powered craft operated under contract to the Air Ministry, which carried it to another wharf on the far side of the river. We then transported it across a rough track to the reconditioned huts on the chassis of a Model T Ford and an ancient fire-engine which must have remained from World War I. Wilkins's first task was arranging for the erection of masts and feeders; this he did with the willing help of Joe Airey, the head of the workshop at Slough, who had come up to Orford with Alec Muir, one of the most versatile of his staff. Under Wilkins's instructions, they performed wonders with the antennae and feeders on the 75 foot towers which had been specially built for the project.

Wilkins and Bainbridge-Bell assembled the receiver and its cathode-ray indicator and I assembled the transmitter. Unlike the

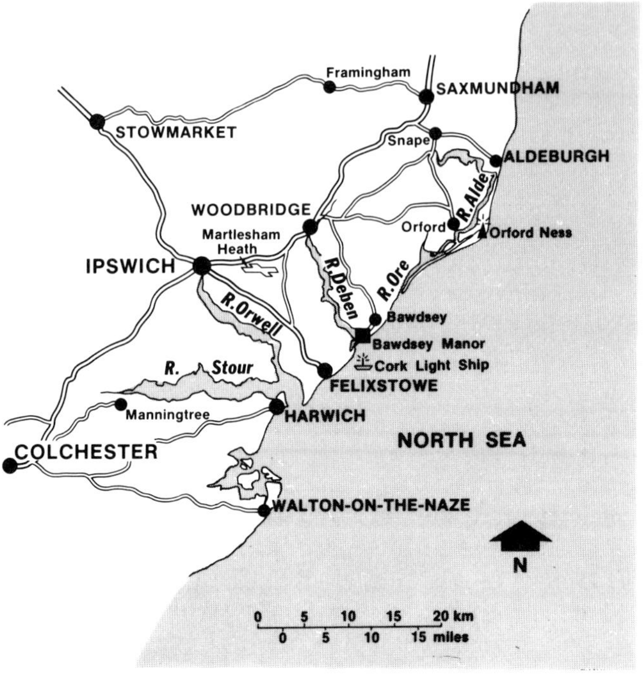

Figure 1.3 A map showing the Orfordness, Bawdsey and Martlesham area as it was in 1935.

receiver, the transmitter was an unknown quantity and had never previously been put together, not even for a test run at Slough. Many of the components were purloined from discarded equipment at Slough and Teddington, in particular some massive high voltage transformers and rectifiers previously used in x-ray equipment. These had scared the living daylights out of Bainbridge-Bell but they were something I was familiar with from my previous research on x-rays at Swansea University. The transmitting valves, at least, were special. These were Naval silica valves type NT46 which were the highest-power transmitting valves used by the British Navy. Although designed for continuous wave operation, by a happy accident they were ideally suited to pulsed operation. They had a tough silica envelope and the electrodes were of molybdenum and capable of running at a dull red heat. The tungsten filaments were unlike anything I had ever seen before and drew about 20 amps of current at 20 volts. The external leads were carried on stalks six or eight inches long and well separated from each other.

We had been carefully instructed on the use of these valves by the staff of the Signal School at Portsmouth; they insisted that the absolute limit of anode potential to be applied was 5000 volts. It was our intention to run the valves near this limit for a start and then to increase first the anode voltage, then the filament voltage until they showed signs of distress. There was not much difficulty in getting the pulsewidth down to 25 microseconds. As a minor part in the reduction of pulsewidth, I made the surprising discovery that a Ferranti audio transformer, specially designed by that company for use in high power audio work, was extraordinarily effective in passing a 25 microsecond pulse with little or no distortion. This was probably the first example of a pulse transformer; later in the war, pulse transformers of much higher power were designed, capable of feeding megawatts directly into an oscillator.

The Orford transmitter was basically a couple of NT46 valves in a push–pull arrangement. With 5000 volts on the anodes and normal filament voltage, they gave a very healthy output of 20 to 25 kilowatts on our first operating wavelength of 50 metres. During the next few weeks I cautiously pushed the anode voltage up to 10000 volts without signs of distress. This was about double the rated value and the output was then around 100 kilowatts, with even more in sight. The transmitter fed a single dipole antenna suspended from the 75 foot masts and we had taken the precaution of putting three nine inch insulators at each end of the radiating element. Soon the power in the antenna was such that it arced over the full 27 inches. Wilkins has described how, on a quiet night, this could be heard over a mile away on the wharf at Orford. The cure was simply to add two copper balls of the type found in lavatory cisterns, one at each end of the antenna.

Later that summer, as the technology improved and our confidence grew, the anode potential was pushed up to 12000 volts. This was the limit reached in my time and the power was then a very healthy 200 kilowatts. This could not compare with the megawatt or so which became available later on in the war, but it was a very satisfactory start for an experimental system. Gradually, order emerged from the chaos of mid-May. We were a small and exceptionally devoted band, feeling our way through uncharted waters, but with the end objective always clearly in sight.

Up to that time we had not received echoes from aircraft, but at the wavelength of 50 metres on which we were operating we were getting an impressive display of echoes from the ionosphere. One surprise discovery was that, in addition to the usual ionospheric echoes at a range of 60 miles, there were extremely strong echoes stretching out to several thousand miles. After a little thought it became clear that these were not from directly overhead but were land echoes from continental Europe, reflected at glancing incidence from the ionosphere—the first example of over-the-horizon radar. With the broad beam we were using at that time, and in the absence of modern signal-processing methods, there was no way on earth of sorting these out. We were, in any case, much more interested in the first 60 miles at that stage and regarded the long-range signals as a nuisance.

In those early days at Orfordness, Wilkins fell naturally into the role of leader of the group. Watson Watt spent most of his time at Slough and in London and his confidence in Wilkins was such that he became the natural line of communication with the group at Orfordness. Wilkins was of a shy and modest disposition; at the same time he was a shrewd judge of his fellow men. He had a wry sense of humour and a facility for inventing nick-names, several of which were bestowed on Bainbridge-Bell. No one could have wished for a better companion than Wilkins, and in a world where judgements are sometimes harsh, I have never heard anyone say a hard word about him.

Bainbridge-Bell was an entirely different character. He was undoubtedly the most talented circuit designer at Slough and was largely responsible for the final design of the cathode-ray direction-finder, which was very advanced for its time. However, at the time of the transfer to Orfordness he had become cranky and a trifle cynical. He did not share the unbounded optimism of Watson Watt, Wilkins and myself and was very pessimistic about the success of the venture. He saw difficulties which simply did not exist. For example, he was quite certain that the increase in transmitter power would lead to unacceptable overload of the receiver and that this would lead to all manner of problems. Almost as his price for joining the project, he had insisted on a very expensive high-frequency cable being laid between the transmitting and receiving huts at Orfordness so that a portion of the transmitted pulse could be fed into the receiver in antiphase to overcome the overload. In the event, the problem simply did not arise and the cable lies buried and unused at Orfordness to this day. This, and his general pessimism, did not go down well with Watson Watt who soon found it expedient to move him back to Slough. There, Bainbridge-Bell was much more in his element and in surroundings which he found far more congenial than the austerity and the grind of Orfordness.

Willis, who had been Bainbridge-Bell's assistant for many years, stayed on at Orfordness and gave every indication of enjoying a release from the rather close rein which had previously been imposed on him. Carter, a very expert and exceptionally hard-working technical assistant, joined the group in place of Bainbridge-Bell and it was this exceedingly small team of Wilkins, Willis, Carter and myself who saw the project through its initial stages.

Success came just over a month after our arrival at Orfordness. On the weekend of 15 June 1935, the Tizard Committee visited the island to review progress. They had not been led to expect a demonstration of aircraft detection, but they examined our equipment with interest and were reasonably impressed with what they saw. Unknown to us, however, Watson Watt, in his optimistic fashion, had arranged for a Vickers Valentia to fly short legs between Martlesham and Orfordness during the Committee's visit. No clearly recognisable signals were received from the Valentia. Many years later Watson Watt was inclined to say that he saw aircraft echoes on that occasion but I am afraid that this was not the case; or at least, if he saw such signals, no one else did. What the Tizard Committee did see was a spectacular display of ionospheric echoes at 60 miles and beyond, but no aircraft echoes. They appeared well satisfied with this. Clearly, much progress had been made and the station was at least clean and tidy—we had bought a mop and pail at the local ironmonger's store and had a good clean-up before they arrived.

I can well remember a fairly general discussion taking place during the tea break, when for the first time I heard mention of Professor Lindemann, the controversial Professor of Physics at Oxford University. At that time both Oxford and Cambridge University were represented by members in Parliament. Professor A V Hill, already on the Tizard Committee, was the member for Cambridge. Lindemann had just put his name forward as a Conservative candidate for the Oxford seat, but his candidature was turned down by the Party. It may be hindsight, but I seem to remember a distinct feeling of satisfaction being expressed at the time about this set-back for Lindemann. Had we but known, there were to be ominous developments later.

Before the Committee departed there was an amusing incident which showed that Rowe, the secretary of the Committee, was not without a sense of humour. It was already late in the day when someone remarked on the fact that there was a spectacular sunset. The Committee mounted one of the adjoining dykes to get a better view. This was an opportunity too good to be missed and Rowe pointed out that if seen from between one's legs a sunset looked very much better. To a man, the members of the Committee turned their backs on the setting sun, put their heads between their knees and remained thus for a few minutes judging the result. Luckily there was no one else within miles to see four such eminent men lined up in that hilarious posture. On the whole, they were inclined to agree that they had not been misled and that the colours really did appear more vivid when viewed upside-down!

The next day, Monday 17 June, Watson Watt took the unusual step of staying at Orfordness for an extra day. No equipment changes had been made from the day before, but when the gear was run up a clear echo immediately appeared at 17 miles. This turned out to be from a Scapa flying boat from the Felixstowe Air Station flying off the coast at Orfordness. Good echoes were received as it moved up and down the coast, after which it returned to base about a half an hour later. We were all wildly excited. My recollection of subsequent events is not clear, but I remember that Watson Watt rang up the Commanding Officer at Felixstowe and asked if one of his flying boats had just landed. 'Yes' said the Commanding Officer and on receiving this assurance Watson Watt promptly asked: 'Would you please ask the pilot to take off and fly the same track again!' This he did and we had a further hour or so of unadulterated pleasure as we followed his passage up and down the coast.

From that moment, progress was very rapid. Martlesham assumed responsibility for test flights of the new radar system and a series of flights were made by aircraft flying on the landward side from Orfordness to Bircham Newton and back, usually at a height of 15000 feet. With such flights on a daily basis, the performance of the equipment rapidly improved; the maximum range of detection increased to 40 miles in September, and to 80 miles by the end of the year. It reached more than 100 miles in the early part of 1936.

Several of us lived in 'digs' at Orford and I was lucky enough to find a room with Emma Good, just off the main square. She was a delightful spinster who looked after me as I had seldom been looked after before. I thought at the time that she must have been about 80 years of age, but this was far from being the case because she lived for at least another 30 years. Wilkins, Willis and Carter settled for the more opulent streets of Aldeburgh and motored in to Orford each day. There was a recognised time for starting at the quay each day, about 8.30 AM, but the finishing time was very flexible, depending on the work on hand. Many were the nights I spent on a camp bed alongside the transmitter, with a piece of cake and a bottle of beer for supper, and as likely as not, the same for breakfast.

We soon became known as the 'Islanders', although Orfordness was not an island and we were hardly of nautical bent. Rowe has variously described Orfordness as 'surely one of the loveliest places in the world' with 'pink thrift and yellow shingle and the cries of the terns'. But his views are heavily biased, and to us Orfordness was a very forbidding place—a land of freezing winds, shingle, mud flats and dykes and a few comfortless huts. By contrast, the village of Orford was a delightful spot, with a castle dating from the time of Henry II, an hotel of some standing called the Crown and Castle and a most hospitable pub bearing a sign 'The Jolly Sailor'.

Orford itself had a population of the most charming people it was possible to imagine. First of all there were the Brinkleys—Tom,

Figure 1.4 The Crown and Castle Hotel, Orford, where the pioneers had many late night discussions on the technology of air warning. (Courtesy of Trust House Forte.)

Vic and George—who were highly accomplished watermen and dominated the river. Then there was Steve Harper, the landlord of The Jolly Sailor, who had the distinction of being painted by Dugdale, a painting which was designated the Royal Academy portrait of the year in 1934. Willie George Sharman, a man of enormous versatility, was at once the local insurance agent and the Secretary *and* Treasurer of both the football and the cricket clubs. Several days a week he was also the local hairdresser and barber. He held court in a room alongside his home, where the cost of a passable haircut was ten pence and a shave sixpence—half price if you were prepared to apply your own lather. Along with The Jolly Sailor this was the local club, where in the late afternoon the workers of the community gathered after their labours and caught up on the local gossip. In addition to this, Willie George ran a most important messenger service. Twice a week he drove into Woodbridge and Ipswich in a van of ancient lineage and bought every kind of commodity for the local housewives—reels of cotton, skeins of wool carefully matched to the samples provided—anything ranging from feminine knick-knacks to bits of ironmongery not available in Orford itself. He was the much loved universal provider of the Orford community, with a nice line in gossip to go with it.

Figure 1.5 Steve Harper, the landlord of The Jolly Sailor painted by T C Dugdale, ARA.

Another great character was Wally Green, the custodian of Orford Castle. Many years later, while on a short visit to Orford, I stayed overnight at the Crown and Castle. Next morning I was served breakfast by a delightful lass who was obviously of local origin. I may have been suffering an excess of nostalgia but I swear that she smelled of new mown hay. I waited for a suitable moment and asked 'Is your name Green or is it Brinkley?' She said 'Green', I then said: 'I knew your father, Wally Green. He was the keeper at the Castle and used to set the fastest spinnaker in the Orford Regatta.' But apparently, I was wrong; it was Jack not Wally Green. Then with a look which owed something to the Mona Lisa, she said 'But he's not my father, he's my grandpa.'

They were idyllic times. In spite of the crude conditions under which we lived and the lack of amenities, for me it was one of the happiest periods of my life. It was a time of high achievement and very obvious progress and we did not care a jot about the absence of refinement in our private lives. The work itself was enormously satisfying but there were also plenty of side benefits. Orford had a clearly defined local aristocracy. The neighbourhood doctor was Dr Horatio Nelson Baron who appeared in the telephone directory as Baron Horatio Nelson. This confused us for quite some time but apparently he really was distantly related to the great seafarer. He had a delightful and most talented daughter Joyce Baron who sailed a 14 foot International—this was the crack sailing dinghy of that era. Her boat was called 'Plain Jane' and in the weekend races on the river she usually defeated all comers.

We soon found that Orford was a haven for retired people, many of whom had been tea-planters and rubber-planters in the Far East. There was a delightful sprinkling of young ladies whom we wrongly called the tea-planters' daughters. I especially remember Betty Johnston, a vivacious girl who sailed another 14 footer with distinction. She had a brother, Stephen, who was an instructor in the RAF and finished the war as a Mosquito pilot in the Pathfinder force. Then there was April Beswick, an ethereal character who reminded me of someone out of J M Barrie. Many were the weekends I spent with them at tennis or sailing on the glorious River Ore. I am inclined to go along with Rowe about the delights of the area, but with an all-important reservation: it was the village of Orford, not Orfordness, which was the seductive paradise.

In those early days, it was Watson Watt's habit to come up to Orford for the weekend to review progress on the island. He and his wife Margaret would stay at the Crown and Castle Hotel and either on the Friday or the Saturday night he would entertain Wilkins and myself to dinner, together with any official visitors, of whom there were an increasing number. After dinner, we would have a wide-ranging discussion about future plans. This was liable to go on far into the night and would cover the design and construction of a chain of five stations to protect the Thames estuary, a new proposal then under serious consideration by the Air Ministry; the problem of distinguishing friendly aircraft from those of the enemy; the problems of searchlight control and gun-laying which Watson Watt had started to discuss with the Army; and lastly, the next task likely to face the Air Force, namely the interception of night bombers—an altogether more difficult problem than intercepting day bombers.

On their next visit a few months later, the Tizard Committee was treated to a full demonstration of air detection with an aircraft tracked to a distance of 60 miles. However, there was a sour note, and in this respect the occasion was less than happy. Due to his insistence on something being done about air defence, Winston Churchill had been added to the influential Imperial Defence Committee; much to the chagrin of the other members of the Committee, he had insisted that his friend Professor Lindemann be added to the Tizard Committee. Unlike the other members of the Committee, Lindemann was not in the slightest impressed with what he saw at Orfordness and he found fault with almost everything—especially the primitive conditions under which we worked and the picnic lunch of cheese, salad and local beer which had been specially laid on from The Jolly Sailor. This latter was much appreciated by the other members of the Committee, but it did not match up to Lindemann's fastidious tastes and he made his displeasure very obvious.

Within a few months of his joining the Committee, what had previously been an innovative and forward-looking group became riven with strife. It was strictly Lindemann versus the rest, with his hostility to radar and his insistence on totally impracticable ideas about intercepting enemy aircraft by means of wires dangled from balloons, or by infrared, which at that time simply did not have the sensitivity to detect aircraft at long range. The situation within the Committee deteriorated so badly that it could not continue and after some monumental rows, including a threat of resignation by most of the original members, Lord Swinton, the Secretary of State for Air, had to step in. He solved the problem by dissolving the Committee and reforming it soon afterwards with Professor Appleton as the new member in place of Lindemann. For a short time at least, peace reigned and productivity was restored, but unfortunately this did not last much beyond the outbreak of war.

Work at Orfordness progressed. The original wavelength of 50 metres had been chosen on the assumption that the best reflected signal from an aircraft would come on the wavelength at which the wing span acted as a dipole resonator. Many bombers of that period had a wing span of about 75 feet, and thus would resonate at a wavelength of 50 metres; and so the first operating wavelength at Orfordness was chosen. However, due to massive interference which took the form of long-distance commercial traffic coming in via the ionosphere, we were soon forced to abandon the 50 metre wavelength and changed to 26 metres without loss of performance. It was not long before this band also became untenable and a further change was made, first to 13 metres and finally to the range 10 to 13 metres. This was retained as standard by the British air-warning system until the end of the war.

As pointed out in Watson Watt's memorandum, the design of a complete warning system required not only range information, but bearing and height as well. The next important step at Orfordness was the addition of a height-finding facility by Wilkins, using a method he had pioneered at Slough for measuring the downcoming angle of transatlantic signals. A little later, an accurate method of direction-finding using crossed dipoles was added and the promise of Watson Watt's memorandum was met.

This success prompted the Air Staff to ask for a set of five stations to provide air warning over the approaches to the Thames estuary and, to their eternal credit, the Treasury earmarked a sum of £1000000 for this purpose. The date was December 1935 and, as we now know, this was the step which consolidated Britain's lead in the technology of air defence. There can be no doubt that within the Government the man most responsible for bringing this about was Lord Swinton, the energetic Secretary of State for Air, who was then busily engaged in expanding the Royal Air Force and giving them the facilities needed to counter the coming threat. In this, he was supported wholeheartedly by the Air Staff, by the Ministry and by Tizard and his influential colleagues. Outside the Government, there is no doubt that the constant lobbying by Winston Churchill for some action on air defence was an important supporting factor.

The task of designing and building a chain of five stations around the estuary was well beyond the capacity of the minute staff at Orfordness. Some thought was given to increasing the staff and finding a new and more favourable site at which to work. On one of his weekend visits, Watson Watt, Wilkins and I went searching for possible sites in the Orford area—in a brand new Daimler with semi-automatic gears which Watson Watt had just acquired. We headed north to the relatively affluent Aldeburgh region but could not find anything remotely resembling a good site for radar research. Then Wilkins had a brain-wave. A couple of weeks before, he had driven down to Bawdsey, an area we had not previously explored. He reported that it had the unusual feature of some high ground—all of 70 or 80 feet above sea-level which was unusual for that part of the coast—and that it was an exceedingly pleasant place with a Manor and an estate which might conceivably be for sale.

We drove down at once and viewed the exterior. It was an entrancing spot and would make an ideal place for a radar establishment. It did not take long to establish that all the requirements were met. It belonged to Sir Cuthbert Quilter and it was for sale, and within an exceedingly short space of time the purchase was negotiated—the Manor together with about 180 acres of prime land—for a total of £23000. I believe it is true to say that the building and land are worth many millions of pounds today. It belongs to the British taxpayer and is recognised by the National Trust as part of the British national heritage; I still get an inner glow from the fact that I played a small part in choosing it.

Bawdsey Manor

The move to Bawdsey began in March 1936 and gradually we took over the Manor and the estate. The establishment was called the Bawdsey Research Station with Watson Watt as the first Superintendent. Laboratories were set up in the White Tower and in the Stable Block and 240 foot towers began to appear on the hill behind the Manor. Up to that time the staff had been employed by the National Physical Laboratory, but on 1 August 1936 they transferred entirely to Air Ministry employment.

Watson Watt took up residence in a palatial flat overlooking the water and at his suggestion batchelor quarters and a Mess were established in the Manor for the single members of the staff. Those who were married—and there were not many—lived in the salubrious township of Felixstowe and came to work each day by ferry across the River Deben. In a very short space of time, Bawdsey Manor became an exceedingly busy place with all the attributes of a thriving research establishment. The staff was growing fast, optimism was high and construction of the new stations around the Thames estuary was proceeding well. It was a delightful existence. We worked hard and could often be found at work well after midnight. At the same time we were not averse to having a swim before lunch or a game of cricket on the magnificent lawn in front of the Manor before dinner. The atmosphere has been compared with that of an Oxford or Cambridge college, where some of the best technical discussions took place late at night in a timbered hall in front of a roaring fire. Among the amenities was a billiard room at the eastern end of the Manor constructed in the 1890s by the original Sir Cuthbert Quilter. The walls of this room were beautifully timbered and it contained a full sized billiard table of high quality and considerable antiquity. When the Air Ministry bought the property towards the end of 1935, they decided that the billiard table could not be included in the purchase price. Sir Cuthbert was reluctant to remove it from the Manor, so as a friendly gesture, he offered it to us for the sum of £25. I knew a bargain when I saw one and

Figure 1.6 Bawdsey Manor at the mouth of the River Deben. The headquarters of radar research in Britain from 1936 until the outbreak of war in 1939. (Courtesy of Cambridge University Press.)

promptly bought it on behalf of the Bawdsey Mess, which was then being formed.

Many years later, there was an interesting outcome. On 15 June 1985, I was honoured by an invitation to attend a Reunion of the Bawdsey Association, an event which takes place each year. It was full of memories for me; it was 50 years to the day since the Tizard Committee made its historic visit to Orfordness to view the original air-warning equipment. During my 1985 visit I found that the billiard table was still in place and was greatly appreciated by the present staff of the establishment. I realised that it was still owned by the original radar team and, on their behalf, I bequeathed the billiard table to the RAF Mess, Bawdsey Manor, for the rest of time. It now carries a brass plate commemorating this fact.

I cannot resist adding another amusing item. In 1975 the Air Ministry decided to close down the Bawdsey establishment and, as a first step, the whole of the interior furnishings were put up for sale. A long list of contents was drawn up and a public auction arranged; among the items for sale was our billiard table. A wide-awake Warrant Officer spotted that an outrageous violation of private ownership was about to be perpetrated; he fixed it in the simplest possible way, by crossing the billiard table off the list. Thanks to that Warrant Officer, the billiard table is still in place; it has been there for almost a hundred years, and I like to think that it will be there for another hundred.

By September 1936 things at Bawdsey had progressed to a stage at which a fairly ambitious air defence exercise was attempted in the Thames estuary. Construction of the five stations had not proceeded far enough for them to participate, so the exercise was focused on Bawdsey itself. About a hundred aircraft were involved over a three-day period. Half of these were bombers which were directed to make simulated raids on Bawdsey from about 100 miles to seaward. The other half were fighters which, under direction from Bawdsey, would attempt to intercept them.

Disaster struck on the first day when, for reasons which were not immediately apparent, the Bawdsey station failed to receive signals from the incoming aircraft at a range sufficient for any interceptions to be made. This was doubly embarrassing because the Commander-in-Chief of Fighter Command, Air Vice-Marshal Sir Hugh Dowding, had been invited to view the operation. He sat in the receiver room and saw nothing until his ears told him that the raiding aircraft were passing overhead. As described in later chapters, I was not directly involved in the exercise and was spending all my time preparing for airborne radar. In preparation for such tests, I had built a 6.8 metre transmitter on the Orfordness model in the Red Tower at the Manor. This, combined with a miniature receiver and indicator suitable for use in an aircraft, gave a range of detection from the ground of close to 50 miles. On the morning of the exercise it happened that I was in the Red Tower making adjustments to this set. I had never previously seen such a glorious display of echoes. For most of the morning, the screen of my receiver was filled with aircraft echoes right out to the limit of range.

I did not hear of the débâcle at the main station until lunch-time, when I discovered the rest of the staff in suicidal mood. They had only been able to detect the incoming aircraft at a range of 10 to 15 miles and no directions had been given to the fighters. Watson Watt was furious. In an effort to cheer him up, I told him about the display on the 6.8 metre set that morning. He insisted on Dowding seeing it at once, so we climbed back up the Red Tower. Sure enough, there was an impressive display of echoes all the way out to 50 miles as the aircraft repositioned themselves for the afternoon exercise. This cheered them up immensely and Watson Watt was kind enough to thank me in Dowding's presence for putting a different complexion on the proceedings.

Unfortunately, the afternoon performance of the Bawdsey station was no better. Someone had blundered in a big way and later that night Watson Watt gathered together a very despondent group to hold the inevitable inquest. There was proof from the Red Tower that the raiding aircraft could be detected out to the limit of range, but the main station had seen practically nothing. A very confused picture emerged from the subsequent discussion. Some put the blame on the new transmitter on the 240 foot tower on the hill, but there were others who defended it. The issue should have been clear cut, but there was a complicating factor. Wilkins had recently replaced the open wire feeders on the receiving towers with coaxial cable and there had not been sufficient time for proper tests to be made. In his modest way, Wilkins was inclined to believe that this might be the problem.

Although I had relinquished responsibility for the main transmitter some time previously, I had grave doubts about the new transmitter on the hill. During the afternoon, I went up to the transmitting hut with a Colebrook wavemeter. This had a neon indicator which lit up when brought into tune with an adjacent transmitter. It was the self-same instrument we had used at Orfordness a year before. The signal which radiated from the old Orford transmitter was such that we could carry the wavemeter half way to the wharf (that is, about a half mile from the transmitter) and still get a healthy glow on the neon indicator. On the afternoon of the exercise, I could get no response at all right alongside the transmitting hut at Bawdsey. I had to climb the transmitting tower to within twenty feet of the aerials before the neon responded—which it did with a very watery signal. Clearly the transmitter was putting out virtually no signal; there was not even a monitor to indicate this.

Our old transmitter was still at Orfordness, but in a very run-down condition. No one had thought it necessary to maintain it—not even as a reference to compare with the new transmitter at Bawdsey. Watson Watt asked if it could be brought back on the air and, as a result, A G Touch and myself were elected to run it up. Touch was a new recruit from Oxford University of whom we shall hear more later. He too had no direct responsibility for the exercise and was acting as my number two on airborne radar. Later that night, he and I drove to Orford with a few tools, some spares and a packet of sandwiches to sustain us. We rowed across to the island, walked the half mile or so to the transmitting hut and started work on the transmitter. It was in a sadly depleted condition but after some massive surgery we had it running again by day-break. As far as I can recall, we fed the output into a single dipole antenna on the 75 foot masts, as in the original set-up, and the performance was close to that of a year before.

The Bawdsey receiver was run up next morning using the Orfordness transmitter and it was clear that something comparable with the original performance was restored. Later that afternoon, the Bawdsey transmitter came back on the air and, operating into a four-element array on the 240 foot mast, the performance was judged to be about the same as that of the Orford transmitter. The remainder of the exercise went off without incident, a high percentage of interceptions was achieved and, apart from the trauma of that disastrous first day, it was judged to be a success. In fairness to some of the very dedicated people involved, and particularly in fairness to Wilkins, who was unjustly blamed for what happened, it is probably true to say that the exercise was laid on much too early. A great many design changes were being made in preparation for the construction of the chain and the mistake was to introduce too many of them at once. It could have happened to any one of us.

As construction of the estuary stations proceeded, another vital component was added to the scheme. The Tizard Committee, alert as always to the problems in hand, pointed out that it was not sufficient just to get warning of approach of enemy aircraft—there was a need to coordinate the information coming from the different stations, for decisions to be taken on which squadrons should be deployed against the enemy, and for precise instructions to be given to the fighters. A test procedure for such a coordinated defence system was worked out at Biggin Hill—it became known as the Biggin Hill experiment. This proved a resounding success and the lessons learned meant that when the final air defence system came into being it was a fully integrated one. It consisted of the air-warning network, a filtering process for assessing and collating the data and an efficient communications system for alerting the fighters and guiding them to their target.

An Air Force group under the leadership of Squadron Leader Raymund Hart was attached to Bawdsey to coordinate this work and the chain network became very much a joint venture, pursued by the scientists in closest cooperation with the RAF. When this organisation was in place, another exercise was held in the autumn of 1937. All five stations around the estuary and the filter system were involved and it was a resounding success. This was the basic system which laid the foundation for success during the forthcoming Battle of Britain.

The five stations were working when Chamberlain made his infamous journey to Munich in 1938 to appease Adolf Hitler. He probably did not know it in explicit terms, but his aircraft was tracked on the way out and again on its way back, to a range of more than 100 miles by the stations around the Thames estuary. The performance of these stations was so good that even before the end of 1936 plans were made to extend the air-warning network to a chain of 19 stations along the whole of the east coast. Wilkins was in charge of site selection and overall design of the system. A specification of the characteristics of a desirable site was drawn up and, like all such documents, it was a rather boring piece of paper. It detailed the preferred height of the site

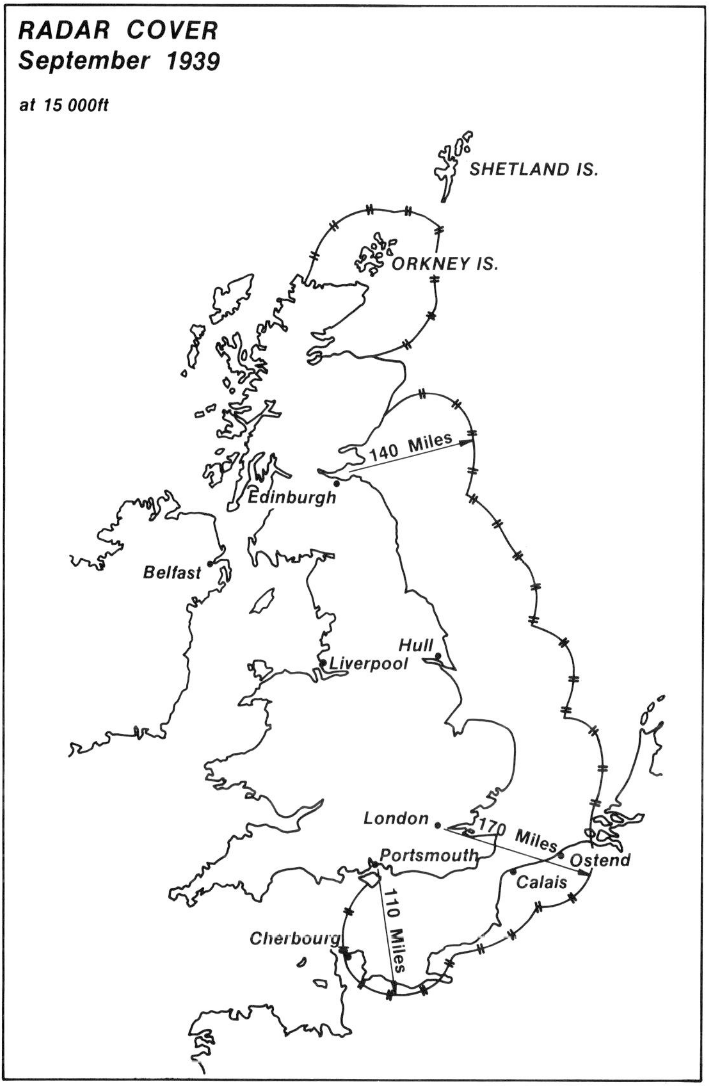

Figure 1.7 The air-warning coverage provided by the east- and south-coast chains against aircraft at 15000 feet at the outbreak of war in September 1939.

above sea level, its distance from the coast, the availability of adequate electrical power and so on. Naturally, this was vetted by the Air Ministry before it was approved. It came back intact except for one amendment—that the choice of site '. . . should not gravely interfere with the grouse shooting'. The additional stations were being built on the Yorkshire moors and in Scotland and someone was making quite certain that, in planning the Air Defence of Great Britain, the right priorities would be observed.

The Metropolitan Vickers Company were given responsibility for producing the chain transmitters, A C Cossor for the receivers and the Marconi Company for the aerial arrays. Construction of the system proceeded apace. The original 19 stations of the east coast chain were later extended by 6 more stations to cover the south coast. The total cost of contracts placed before the outbreak of war was some £10 million, which was a measure of the importance invested in the endeavour by the British Government. A map of the total coverage by these stations at the outbreak of war in 1939 is given in figure 1.7.

This was the air-warning network which became the crucial element in the great air battle which was fought over Britain in the autumn of 1940. The story has been told many times. The Spitfires and Hurricanes of the day were superior to the aircraft against them, but they were few in number. The pilots and men of Fighter Command put up a magnificent defence but their endurance was strained to the limit before the battle ended. The chain system allowed them to stay on the ground until the last possible moment and then steered them unerringly to their targets. This so enabled them to conserve their efforts that a numerically inferior force was able to beat back an enemy which would otherwise have overwhelmed them.

In his final account of the Battle of Britain, the Commander-in-Chief of Fighter Command, in typically conservative fashion, said of the radar air-warning system: '. . . it is not too much to say that the warnings it gave could have been obtained by no other means and constituted a vital factor in the Air Defence of Great Britain'. At a later date, Lord Douglas, who succeeded Sir Hugh Dowding as Commander-in-Chief, said very simply: 'I think we can say that the Battle of Britain might never have been won. . . if it were not for the radar chain.'

At the end of the war, General Galland, who had overall command of the German attack on Britain in 1940, was much more explicit. He said: 'The British had, from the first an extraordinary advantage, never to be balanced out at any time in the whole war; their radar and fighter-control network. It was for us and for our leadership a freely expressed surprise, and at that a very bitter one, that Britain had at its disposal a close-meshed radar system, obviously carried to the highest level of current technique, which supplied the British Fighter Command with the most complete basis for direction imaginable. . . . We had nothing like it. . .'

The Battle of Britain was the first time in World War II that the tide of battle turned in Britain's favour. It was a triumph for the pilots and the men and women of Fighter Command which will never be forgotten. But it was also a tribute to the genius of Tizard who foresaw the problem, and to Watson Watt who provided the solution.

2

The Early Development of Airborne Radar

In the early part of 1936, when the success of the chain air-warning system against day bombers was assured, the much more difficult question of intercepting night bombers came up once again. It had never been far from our minds and after the first successes at Orfordness it was placed on the growing list of projects which would be attacked when staff became available.

In later years I saw a good deal more of Tizard than I did at that time and it became clear that it was Tizard and the members of his Committee who were thinking ahead and providing the leads. Tizard was confident that the air-warning range of nearly 100 miles which had already been achieved by the chain system, when married to the filter and interception procedures which came out of the Biggin Hill experiment, had an excellent chance of repelling the daylight attack on Britain. The German Air Force, he argued, would then turn to night bombing. What could be done about this? This was how Tizard's mind worked and he was to provide many examples of a similar kind over the years. His prediction that the German Air Force would be beaten back in daylight, and that it would then turn to night bombing must rank as one of the best examples of technological forecasting made in the twentieth century.

Reference to the actual figures shows how difficult the problem of night interception was going to be. At night, the range at which an enemy aircraft could be seen and identified was about 1000 feet, and it was clearly beyond the capability of a ground system to place a fighter as close as that. What was required was a radar small enough to be installed in a night fighter, which would enable a pilot on his own initiative to close from a range of 4 or 5 miles down to the required minimum of 500 or 1000 feet. In the light of many discussions of night interception which I had with Tizard in later years, it became quite clear that he was probably the one man who, in the mid-1930s, correctly foresaw how the situation would develop and what should be done about it. He passed these thoughts on to Watson Watt and Watson Watt passed them on to us at Orfordness.

No one was very optimistic about being able to make such a miniature radar at that time. The existing transmitter at Orfordness was a whole room full of equipment weighing many tons. The transmitting and receiving antennae were on masts 75 feet high, soon to be increased to huge structures 240 feet high. The receiver was a large rack of equipment bristling with valves, control knobs and indicators requiring the attention of a highly skilled operator. Miniaturisation of the kind common in the 1950s was still unknown and the marvels of solid state technology were not even dreamed about. There were other formidable obstacles. To achieve a reasonable antenna size, the operating wavelength would have to be reduced to one or two metres. This was at a time when the shortest wavelength for which components were available was five metres. In addition, in order to achieve a minimum range of 1000 feet, the pulsewidth had to be reduced to near one microsecond, that is, to one millionth of a second. The reduction in pulsewidth from 200 microseconds to 20 microseconds which had already been achieved at Orfordness had not been difficult, but going down to 1 microsecond was strictly unknown territory.

The move to Bawdsey Manor was complete by mid-1936. The staff numbers were still small and a decision had to be made about the future activities of the senior members of the group. There were two directions in which to go. One was to put an all-out effort into the construction of the chain air-warning system, the design of which was rapidly taking shape. The other was to put some effort, however small, into new ventures of which airborne radar was one. To be quite frank, I was not attracted to the demanding business of selecting sites for the fledgling east-coast chain and to the long slog of discussions with contractors on the final design of the ground stations. Much more attractive to me was the new and exciting field of airborne radar, difficult though it might prove to be. What is more, I had just had my first flight in an Air Force aircraft, a Westland Wallace with two-stage blowers—it was literally the first aircraft to fly over Everest in 1933. After a long discussion at the Crown and Castle which went far into one Friday night, I pressed hard for us to embark on the airborne project. Watson Watt was in favour and decided that, commencing with our arrival at Bawdsey, Wilkins would take broad responsibility for the construction of the east-coast chain and I would become the leader of a new venture, the airborne radar group with a staff of one—myself. It is typical of the Scottish caution with which Watson Watt approached decisions of this kind that he made it a part-time assignment only. In the absence of any one else, I was still responsible for the main transmitter at Orfordness.

My first step was to draw up a list of the design criteria which should be aimed at in an airborne system. There was no such thing as an operational requirement and no precedents by which mundane considerations like size and weight could be evaluated. A start was made by discussing the whole problem with Fred Rowarth and N E Rowe, the two resident engineers at Martlesham Heath. They had a tremendous reputation as men steeped in the mechanics of what could and could not be done in the air. They were among the few people at that establishment who had been admitted to the secrets of Bawdsey. I then had discussions with such officers of the Air Ministry and Fighter Command as could be induced to talk about the night interception problem. Needless to say, in 1936 there was no section of the Air Ministry or of Fighter Command which was responsible for night interception and there were no aircraft designated as night fighters. As to projected aircraft, there were as many opinions as to what a night fighter would look like as there were people prepared to make a guess.

Gradually, some more or less self-imposed guidelines came together. It looked as if the projected airborne radar would be limited to a total weight of 200 pounds, it could occupy a maximum of 8 cubic feet and should consume no more than 500 watts of electrical power. The total power available on aircraft of the Royal Air Force at the time was just 500 watts at 12 volts, shortly to be increased to a more acceptable 24 volts. Most of this was absorbed by existing services on the aircraft and there was no surplus power available for a radar set. Lastly, everyone was rightly concerned about aerodynamic drag and there was virtual consensus that there should be no long wire antennae of the type then in vogue: the most that could be tolerated were stub antennae, a foot or so in length. This again demanded an operating wavelength of about one metre. These were truly a formidable set of requirements, given the state of radio science in 1936. There was yet another constraint. The equipment would have to be simple enough to be operated either by the pilot alone, or at most by a radar operator functioning in a cramped, noisy and pitch-black environment.

About this time, we stumbled on a gem beyond price; it was a tuned radio-frequency (TRF) receiver designed by the EMI Company for their projected television service from Alexandra Palace. It operated on a frequency of 45 megacycles per second (Mc/s) or a wavelength of

6.7 metres and had a band width of 1 Mc/s. It had 7 or 8 valves on a chassis about 3 inches wide by 15 or 18 inches long. I cannot quote the sensitivity, but it was far and away better than anything which had been achieved in Britain up to that time. This receiver formed the basis of our whole airborne radar experimental programme for the next two years. I have never discovered exactly how we came by that particular set, but I suspect it came through the back door of the EMI Company; they had no idea of the use to which it was being put. During the next few years we made strenuous efforts to obtain additional receivers of the same design; but although the negotiations were conducted by Watson Watt himself, he failed to produce another chassis. This account is being written some 50 years later and it is difficult to believe that until the end of 1938, when we had two Ansons and three Battles as experimental aircraft and at least that number of radar transmitters to share between them, we only had the one receiver chassis which was switched from one aircraft to the other as occasion demanded. Reverting to 1936, when we combined the EMI receiver with a simple cathode-ray tube display system, the airborne receiving problem was virtually solved. The total weight of this combination was less than twenty pounds and the input power minimal. The wavelength was still 6.7 metres, but for the moment this would have to suffice. At that stage there was simply no hope of building a 6.7 metre pulse transmitter which could be carried in an aircraft.

We were now comfortably installed at Bawdsey Manor with laboratory space in the White Tower and roof space on both the White and the Red Towers. I had also been lucky enough to be given some assistants. The first of these was Gerald Touch, a PhD researcher from the Clarendon Laboratory in Oxford, followed by Sidney Jefferson and Perc Hibberd, both with extensive experience in the radio industry. A little later we were joined by Keith Wood, the first of several exceptionally talented technical assistants we recruited from the Felixstowe area. I could not have had better colleagues. Touch and Wood played a prominent part in all aspects of airborne radar until the end of the war, while Jefferson and Hibberd, although destined to move to other spheres, brought much-needed engineering talents to bear on the airborne problem.

It finally occurred to me that we could begin experimental work by leaving the transmitter on the ground, carrying only the receiver and indicator in an aircraft. This we proceeded to do by building a 6.7 metre pulse transmitter on the roof of the Red Tower along the lines of the Orfordness model. The pulsewidth was 3 or 4 microseconds and the power 30 to 40 kilowatts. The first tests were made on the ground with a single-dipole transmitting antenna on the Red Tower and the

Figure 2.1 Dr A G Touch, second-in-charge of the airborne radar team and the architect of ASV Mark II.

Figure 2.2 'Perc' Hibberd who built the first airborne radar transmitter at Bawdsey. Later, at the Air Ministry and the Ministry of Aircraft Production, he played a vital part in supervising the manufacture of airborne radars by British industry.

Figure 2.3 Keith Wood who participated in most of the early flight trials of airborne radar. He played a crucial role in the design of aerial systems for both air-interception and sea-search radars.

receiver in the White Tower. Maximum ranges of 40 or 50 miles on target aircraft were soon achieved. This was the moment to ask for the first airborne tests and these were soon arranged. At this stage the Aeroplane and Armament Experimental Establishment at Martlesham Heath was responsible for routine test flying for the radar chain. The Commanding Officer was Group Captain A C Maund, a man of tremendous personal charm, beloved alike by his staff and by the aircraft industry with whom he was constantly in touch. His cooperation with the new and rather demanding radar group at Bawdsey could not have been more friendly or more willingly given. He arranged for one of his Handley Page Heyfords to be made available for the airborne trials.

One reason for the choice of the Heyford was that the in-line Kestrel engines carried a well designed ignition harness. As a result, the aircraft was reasonably quiet electrically, unlike some of the older, radial-engined machines which produced so much electrical noise that they put a radar receiver completely out of action. The pilot was a Flight Sergeant with the unlikely name of Shippobotham. Among a galaxy of talented pilots of all ranks which graced Martlesham Heath at that time, he was one of three with whom we were closely associated

Figure 2.4 The author, Dr 'Taffy' Bowen, a member of the original radar team at Orfordness and Bawdsey and leader of the airborne radar group. Here, he is seated in the rear cockpit of the Fairey Battle K9208, the principal aircraft used for test flights of 200 Mc/s air-interception radar.

and for whom I have a great affection. The others were Flight Sergeants Wareham and Slee. All three were exceptionally talented and had distinguished careers as test pilots. Slee, for example, became a Wing Commander and a senior member of the King's Flight. He had the distinction of flying King George VI to Gibraltar and Malta during the height of hostilities.

So the first flight trials were conducted in the autumn of 1936 with the relatively massive 6.7 metre transmitter on the ground at Bawdsey and the receiver and indicator in the Heyford. The receiving aerial was a half-wave dipole strung between the wheels of the Heyford and the power came from a miscellaneous collection of accumulators and dry batteries on the floor of the aircraft. We trundled these into the machine before each flight and carried them out again at the end of it. The high-tension supply for the cathode-ray tube came from a second-hand Ford ignition coil driven from a vibrator operated by a 12 volt battery. The first flights took place in the autumn of 1936 out of Martlesham with Flight Sergeant Shippobotham as pilot and myself as observer. On the very first flight, while flying at a height of 2000 or

3000 feet over the Bawdsey transmitter, we obtained echoes from aircraft which were in the circuit area at Martlesham. The range of detection was 8 to 10 miles; this was repeated on several subsequent flights and the maximum range achieved was around 12 miles. It is interesting to note that this range was greater than anything obtained by operational radars of the conventional type during the whole of the war years. A range in excess of 10 miles was certainly never exceeded at such a low flying height.

At that time, the ground air-warning system was known as RDF 1 and the projected airborne system as RDF 2. The new system with the transmitter on the ground and the receiver in the aircraft became known as RDF 1.5. Such schemes are now much used in missile delivery systems, but this was the first of its kind. I became a strong advocate of RDF 1.5 and, realising that it would be several years before a complete RDF 2 system would become available, I pressed for it to be adopted as an interim system; unfortunately Watson Watt would have none of it. To be quite fair, there were difficulties. The range measurement was only correct when the fighter was directly between the ground transmitter and the target aircraft. In other positions, the range would be underestimated. In these days of minicomputers this would be no problem at all, but in those primitive days it seemed to be a formidable obstacle.

I persisted and pointed out that if a grid of ground stations was established, the fighter could always select a transmitter which was behind it and get into the favourable stern chase position. The maximum range was potentially 20 miles, or more even for low-flying aircraft and the system would, therefore, not require a highly accurate ground control system. Watson Watt remained unconvinced and the proposal was allowed to lapse. With hindsight, it is now clear that this was a grave mistake. If RDF 1.5 had been given proper Service trials in the RAF, there would have been several important consequences. In the first place, it would have given them an interim device on which test interceptions could have been carried out at night, two whole years before the outbreak of war. This would have provided pilots and observers with training in the techniques of night interception, something which they did not actually get until war was declared. Lastly, it would have given us experience of the immense problems, still to be faced, of introducing new and sophisticated equipment into Service use.

However, rapid progress was being made in the airborne group, which had grown to half a dozen people. We had acquired some Western Electric 316A valves—'giant acorns' or 'door-knobs' as they were variously called—with a potential for operating down to wavelengths of a metre or so. Hibberd built one of these into a miniature transmitter, initially on 6.7 metres to match the EMI receiver. The pulse recurrence frequency was about 1000 per second, the pulsewidth 2 or 3 microseconds, but the power could not have exceeded a few hundred watts. We added Hibberd's transmitter to the Heyford installation in March of 1937 and once again there was success on the first or second flight. Clear signals were received from prominent objects on the ground, particularly from the wharves and cranes at Harwich, which stood out distinctly from the surrounding terrain. In addition, there were distinct echoes in the distance which could have come from ships approaching Harwich on the seaward side. We could not be certain of this at the time as the Heyford was not equipped for flying over water and it was forbidden to go beyond gliding distance from the coast. The maximum range of these objects was already three to four miles and, with an optimism based on our experiences at Orfordness, we were quite sure that this performance would rapidly be improved upon.

The First Ansons for Airborne Radar

Soon after the success of the Heyford flights, it became clear that we could not continue using Martlesham aircraft for an extensive radar programme. The Martlesham aircraft were fully committed to other projects and we were encouraged to ask for an aircraft specially devoted to airborne radar. The enthusiasm generated by the Heyford flights made this an easy task. We were given a free choice of aircraft and the selection was not difficult to make. Although of limited performance, the Avro Anson was an obvious choice. It was a machine without vices, it had a spacious cabin with ample room for two pilots and two observers, and plenty of bench space to spread experimental equipment around. What was probably more important, it was a standard Coastal Command aircraft and was therefore equipped for operation well out to sea—an important consideration now that our potential targets were ships as well as aircraft. However, there was one snag. In standard form, the ignition system of the radial engines was notoriously worse than that of in-line engines. They were badly screened and tests with an early Anson at Martlesham showed that our airborne receiver was completely saturated by noise from the ignition system. However, Farnborough were exceedingly helpful and promised to fit a better ignition harness which would solve the problem.

On the basis of that promise we selected the Anson, and we have never had reason to regret the choice. Much to our suprise, we were allocated not one but two aircraft. These proved to be the foundation

Figure 2.5 Flying Officer 'Blood-orange' Smith, the first Officer-in-Charge of the Radar flight at Martlesham Heath. He did most of the early test flying of airborne radar and was one of the youngest pilots to earn a DFC in World War II.

on which the whole of British airborne radar was subsequently based and must rank as one of the best investments in the future that the Royal Air Force ever made. With the aircraft came a ground crew to service them and four Sergeant Pilots—Sergeants Smith, Naish, Argent and Newman. As flight commander, Pilot Officer D C Smith, of whom more anon, arrived a few days later. The two Ansons, K6260 and K8758, came on detachment from 220 Squadron at Bircham Newton. They arrived at Martlesham in June 1937 and the first priority was to get them off to Farnborough to be fitted with the magic ignition harnesses. This was likely to take several weeks, so we put an intensive effort into what was now uppermost in our minds—namely getting the wavelength of the airborne system down to one metre. Hibberd and I worked on the transmitter while Touch attended to the receiver.

The transmitter was converted to push–pull operation using two Western Electric 316As. The pulse repetition frequency was still around 1000 per second, but we managed to reduce the pulsewidth to one microsecond and the wavelength to a satisfactory 1.25 metres. This appeared to be the lower limit that could be achieved with those particular valves; at shorter wavelengths the power output fell away alarmingly. We did not have much idea of the pulse power output, but it was certainly less than one kilowatt. Touch rebuilt the receiver to

Figure 2.6 The Handley Page Heyford. This was the aircraft used as a radar target by A F Wilkins in the Daventry experiment in 1935. In 1936, a Heyford from Martlesham Heath became the first aircraft in history to carry a successful airborne radar. (Courtesy of Aeroplane Monthly.)

operate at 1.25 metres by converting it to a super-heterodyne, using acorn valves from RCA in a frequency changer ahead of the one and only EMI chassis. The latter now became the IF (intermediate frequency) amplifier, still on its design frequency of 45 Mc/s. It is interesting to note that 45 Mc/s remained the standard intermediate frequency for all airborne and many other radar systems for the remainder of the war, and probably for a long time afterwards.

Another change was made at this point; it was a minor one but it had important implications for the future. For the Heyford flights we had relied upon a hodgepodge of accumulators and dry batteries as the source of power. For the new equipment in the Ansons we needed six volts, four volts and two volts for filament heating on the transmitter, the receiver and the cathode-ray tube respectively. The receiver required a 240 volt anode potential, the cathode-ray tube about 800 volts and the transmitter rather more than 1000 volts. Obviously there was only one way to go. We settled on an alternating current system using a 240 volt, 50 cycle alternator driven by a DC motor running off the regular aircraft supply. This had the further advantage that we could run our airborne radar sets from the standard mains supply when the aircraft was on the ground. We had two of

Figure 2.7 Avro Anson K8758 aircraft acting as target, photographed from the radar-equipped K6260. Taken over Bawdsey Manor and the estuary of the Deben in the summer of 1937.

these motor generator sets, one for each Anson, and they were to be the sole source of power for Great Britain's airborne radars until the advent of engine-driven alternators in 1939, about which I will have more to say later.

So, in August 1937 a new scenario presented itself. The two Ansons came back from Farnborough adequately screened, the newly upgraded transmitter and receiver/indicator were installed in K6260 while K8758 was used as the target aircraft. The first flight with the 1.25 metre equipment was carried out by Touch and Keith Wood—to my infinite regret I missed it. I had not seen my parents for two years and I was home in South Wales on a short leave. Touch and Wood obtained clear echoes from ships off the coast at Felixstowe at ranges of two to three miles on 17 August 1937. Excitement could not have been higher.

In between flights at Martlesham, it was our practice to run the airborne equipment on the roof of the White Tower at Bawdsey, making alterations and adjustments. There was a ready-made target on the horizon—the Trimley water tower, which gave a very distinctive echo at just over three miles. In the absence of power measuring equipment on metre wavelengths and of a signal generator to measure receiver sensitivity, this was the only measure of system performance we had. Using the amplitude of the echo from the Trimley tower as a bench-mark, we soon found that a small increase in wavelength to 1.5 metres or to 200 Mc/s made a worthwhile improvement in sensitivity. This was the moment when 200 Mc/s radar was born, and this was to remain the most widely used frequency band in airborne, shipborne and ground-based radar throughout the war and it had its origins on the roof of the White Tower of Bawdsey Manor and on test flights in the Anson K6260.

Morale in the airborne group could not have been higher. We had an operating airborne radar, with a range of detection on ships of two or three miles and on ground objects of five or six miles. We were still to receive echoes from our principal objective, namely other aircraft, but we did not have long to wait.

Air-Sea Exercise in the North Sea—4 and 5 September 1937

Only a few days after the successful flight in the Anson K6260 I received a telephone call from Watson Watt who was on one of his frequent visits to London. He said that an exercise was planned, starting on 4 September, during which Coastal Command would search for the British Fleet in the North Sea. Would we care to participate? He could give very few details. An unknown number of vessels would make a passage from the Straits of Dover north to Invergordon. They would follow a zigzag course and, naturally, would try to avoid discovery. A total of 48 aircraft of Coastal Command were scheduled to search for them. About half the aircraft would be Ansons and the remainder would be flying boats operating out of various bases from Felixstowe in the south to Leuchars in the north. The search area was the whole of the North Sea and the exercise was planned to last 48 hours, commencing at first light on Saturday 4 September.

This was an opportunity too good to be missed. It happened that the whole of Martlesham was on stand-down for late summer leave. However, we managed to keep back two ground staff to service the aircraft and one pilot, Sergeant Naish, who as it turned out was a most fortunate choice. He was a newly commissioned short-service appointee to the Air Force, prior to which he had been a Master Mariner in the Merchant Marine. This meant that in addition to being an expert pilot, he was a first class navigator who knew the English Channel and the North Sea like the back of his hand.

Keith Wood and myself were to do the flying, with Sergeant Naish as pilot, and during the previous week we spent many hours on the ground and in the air getting the 200 Mc/s equipment into peak condition. By the afternoon of Friday 3 September we were satisfied that everything was in good order for the exercise. About 4.30 that afternoon we felt we deserved a rest and the three of us were relaxing over a well earned beer in the Sergeants' Mess. We were half way through our first drink when Sergeant Naish said 'Do you know what? If I know anything about the British Navy, they'll be coming out of the Solent right now; then they'll sail up the Channel at five knots.' He gave us a few moments to take this in and then said 'Let's go down and take a look.' We downed what was left of our beer, alerted the ground crew, wheeled the Anson K6260 out of its hangar and took off for the Channel. Naish's guess was dead right; we found a small Naval force in mid-channel heading for the Straits of Dover. It consisted of the battleship *Rodney,* the aircraft-carrier *Courageous* and the cruiser *Southampton* in line astern, with six destroyers in attendance. This was the first time we knew which ships and how many were involved. We did several runs over the group, and against such a huge target we obtained far better echoes than anything we had seen up to that time. We returned to Martlesham just before nightfall, well satisfied that our radar was working near its peak and would give a good account of itself in the morning.

Next day, we were up at the crack of dawn and at first light were on the perimeter at Martlesham, with engines ticking over waiting for the last wisps of morning fog to clear. When we could see the trees at the far end of the field, we took off and headed south-east. We crossed the coast at Felixstowe and then headed due east to begin the search. As we crossed the coast, six flying boats took off from their base at Felixstowe through haze which was noticeably thicker than at Martlesham. We were interested to see that they headed north-east—they were obviously intent on searching further north than we had planned. We continued out to sea, executing a square search at 3000 feet with the radar operating and occasionally giving echoes from small craft below. We had been out for some time when suddenly, about 8.00 AM, there was a larger echo at five or six miles. We closed the range and as we got closer most of the Fleet came into view. It was the *Courageous,* the *Southampton* and the attendant destroyers, but not the *Rodney*. As we approached, all hell broke loose below. Signal lights flashed in all directions, guns were fired—no doubt firing blanks—and aircraft started to take off from the *Courageous*. We were told later that she was carrying 15 Swordfish at the ready and all 15 took off to intercept us. For the very first time we saw aircraft echoes between ourselves and the sea at a range of about one mile.

We did not like the idea of that number of aircraft coming to close quarters and climbed to 9000 feet. If they were mad at us, as well they might be, there is no knowing what they would throw in our direction. So at 9000 feet, and well clear of the Swordfish, we set off in a northerly direction to search for the *Rodney*. We guessed that it would be somewhere ahead of the other vessels, not behind them. However, several hours' search failed to find the battleship and, with fuel running low, we thought about turning for home. We had concentrated so much on the radar search that we had failed to notice that there had been a serious deterioration in the weather. Also, we had weaved about so much that we were quite uncertain of our position. Sergeant Naish told us that he had not been able to see the sea surface for at least an hour and that he would go down to a lower altitude. This we were happy about; we had already learned that the Anson was a wonderfully stable machine and because of ground effect could fly quite happily for hours at 10 or 20 feet above the sea. A favourite trick of the ground crew when an aircraft returned from reconnaissance was to rub a finger along the propeller and taste the salt that had been picked up.

So we went down through the haze, which was now quite thick. We got down to a few hundred feet before we could dimly see the surface of a glassy sea. Sergeant Naish settled down at the accustomed 20 feet and we headed west, still uncertain of our position. Forward visibility was hard to judge but it could not have been greater than a few hundred yards. Quite suddenly we saw a hugh black and white marker buoy sailing past the starboard wing tip. 'The Noord Hinder' said Naish—a navigation mark not far off the Dutch Coast. Without even looking at a chart, this gave him the fix he needed and a compass bearing for home. As we approached the coast, we knew we would get several minutes warning of our approach from the radar but Sergeant Naish decided to play it safe and climbed early. It was even thicker than we thought and we had to climb to over 12000 feet before getting into clear air. We could tell from the radar when we actually crossed the coastline and started to let down, knowing that Martlesham Heath was about eight miles inland. We must have been down to 200 or 300 feet before we broke through the low cloud and to our enormous relief found ourselves over typical Suffolk heath country with bracken and a reassuring mixture of yellow gorse and purple heather below. There was a winding lane just underneath us and we recognised it at once. It was the local 'lovers' lane' where Martlesham men were apt to take their girlfriends of an evening. It was directly in line with the main runway! A 180 degree turn and we were on the final approach to Martlesham, with perhaps 15 minutes fuel to spare.

Watson Watt had given me the telephone number of the officer in charge of the exercise at the Air Ministry, a Wing Commander whose name I have forgotten, and asked me to telephone him with the result of the exercise. This I did at about 10.00 AM. I told him that we had found the Fleet at 8.00 that morning and gave him the names of the vessels and a position. He said: 'Yes, that is correct. An aircraft did intercept the Fleet in that position and at that time. But what squadron is this reporting?' He was astonished to find that it was not a squadron aircraft but a lone aircraft from Martlesham Heath with a Sergeant pilot and two civilians as crew. He had not been told that we would be part of the exercise. He then dropped his own bombshell. 'Did you not get our radio signal?' I had to admit that we did not even have a radio on board. 'You will be interested to know' he said 'that due to deteriorating weather the exercise was cancelled two hours ago.' The aircraft of Coastal Command were all safely back on the ground while we were still out there near the Noord Hinder.

This was something of a landmark in the history of airborne radar. We had found the Fleet under conditions which had grounded Coastal Command, we had detected other aircraft for the first time with a self-contained airborne radar and, simply by returning home in one piece, had demonstrated some of its navigational capabilities. From then on our path was much easier. A P Rowe was never a great admirer of the airborne group, but of this episode he was later moved to say 'This, had they known, was the writing on the wall for the German Submarine Service.' It was exactly one year and 364 days before the outbreak of war.

The North Sea exercise generated so much interest that we were besieged for flight demonstrations in K6260. On 10 September 1937 we gave a demonstration to Dr Talbot Paris, the head of the newly formed Army Radar Group at Bawdsey, and on 14 September gave a similar demonstration to Squadron Leader Raymund Hart, who later played such an important part in introducing the new radar technology to the Royal Air Force. These flights were essentially a repeat of the 4 September exercise in which we used the wharves at Harwich and ships in the Thames estuary as targets. On 18 October we gave a demonstration to Sir Henry Tizard. Apart from the fact that it was his first flight in a radar-equipped aircraft, it was a nostalgic occasion for him. He had actually chosen the site of the aerodrome at Martlesham in 1917 and had been the first Officer-in-Charge of its scientific work. Operating out of that field, he had been awarded an AFC for attacking a German aircraft which was foolish enough to cross his path during a test flight.

We took off in typical autumn weather with little wind and a low ceiling. Although the weather was reasonably clear over land, we found the Thames estuary was fog-bound, with the visibility down to a few hundred yards. During most of the flight we saw neither the sky nor the surface of the sea. However, we had the reassurance of seeing the coast behind us on our radar. The estuary seemed to be full of ships, most of them stationary. We must have done eight or ten runs on ship targets and although we were flying at a height of only a few hundred feet the visibility was such that we did not see a single one of them. As a demonstration of ship detection under bad conditions, it could not have been improved upon. Tizard was delighted and the time came to turn for home. As usual, we had been weaving about and were uncertain of our position. However, the radar showed there was land on the port side and we gingerly went down to take a look. It turned out to be the North Foreland, which fixed our position and gave us an easy course for home. We headed north across the estuary at a few hundred feet and made a landfall at what we dimly saw to be Walton-on-the-Naze. We crept along the coast in our accustomed fashion until we reached the mouth of the Deben and turned gratefully for home. Tizard was very pleased by the whole demonstration, but like the rest of us he was probably relieved when we were safely on the ground again.

It was on this occasion that Tizard mentioned the importance of keeping careful records of every flight we made. He pointed out that it was all too easy to rush off on the next step without recording the reasons why that step was taken. In characteristic fashion, he fixed the thought in our minds by telling one of his stories, of which he had an abundant store. Apparently he was supervising some tests at the Naval Air Station at Gosport in World War I. He was depending on a naval rating in the back seat to record details of the flight and he impressed on him the importance of writing down exactly what happened from take-off to the final touch-down. When they returned from the flight, he found that the first entry read as follows:

9.41 AM Climbing at 48 knots. Passed a seagull, flying backwards.

3

Sea Search and Tales of 'D' Flight

By the end of 1937 our two projects, ship detection and air interception, had been graced with the inevitable acronyms—ASV for Air to Surface Vessel and AI for Air Interception. As ships were a much easier target than aircraft, we concentrated our efforts on ship detection, knowing that any improvement in the detection of ships would automatically benefit the detection of aircraft. In fact, 1938 became the ASV year during which we carried out the greater part of the development of ship detection equipment.

By this time we had been joined by Hanbury Brown, a brilliant young student only 21 years of age. He came from Imperial College and had been induced to join the Bawdsey team by Tizard himself. There were several other important appointments—Ron Taylor and 'Yagi' Walters, both with experience in industrial research. We also recruited two more locally trained assistants from Felixstowe, 'Chalky' White and Mills. Both developed a special rapport with the aircrew at Martlesham and proved of utmost value in our installation and flight-testing work. I had a specially soft spot for Chalky White. About that time, I used to run four circuits of the playing field in front of the Manor before dinner. I was rather proud of my time around the course but Chalky was not in the least impressed. One day he offered to come along for a friendly jog and I do not have to say that he crossed the finishing line well ahead of his boss. Later, he used to play outside me on the wing for the Felixstowe Rugby Union team and the outcome was much the same. When I did not know which way to turn, I only had to pass the ball to Chalky and a try would often result. About that time, we were also greatly strengthened by the attachment of Bill Eastwood, another graduate of London University in the

Figure 3.1 Robert Hanbury Brown and 'Chalky' White who were largely responsible for introducing air-interception radar to the Royal Air Force.

mould of Hanbury Brown. Originally an Admiralty appointee, he was seized on by the Army Group at Bawdsey and placed with us to help with 200 Mc/s technology. He later moved back to the Army Group and applied his knowledge to their special applications.

Although we were always trying something new and continually improving the performance of our radars, life settled into a routine. We made instrumental improvements in our equipment in the laboratory at Bawdsey and these were followed by flight trials at Martlesham Heath. We had several regular radar targets. The first were the wharves and cranes at Harwich, easily identified soon after take-off from Martlesham. Another was the abandoned airship shed at Pulham near Norwich, a huge metal structure which, on the radar, stood out like a beacon in otherwise featureless terrain. Then there was the 'Butter Boat' from Esbjerg in Denmark which became our standard ship target. This vessel operated on a regular 24 hour schedule, arriving at Harwich in the late afternoon and leaving again about nightfall. It could always be found offshore in mid-afternoon and we sometimes went far into the Heligoland Bight to meet it. We used to do runs on the Butter Boat, broadside and end-on, carrying out whatever tests we were engaged upon at the time. I am sure that the skipper and his crew came to know us well. They must have wondered why the same aircraft would be there to meet them day after day at about the same spot.

Most of the flying was shared between Keith Wood, Touch and myself, depending on the particular project in hand. The work proceeded smoothly and we enjoyed the alternation of laboratory work followed by flight tests. We were a closely knit group with a high sense of camaraderie which spilled over into the football and cricket field. I can remember many a hard-fought game, both with and against the Martlesham Heath rugby football and cricket teams. It did not seem to matter whether we played for them or against them. Either way, it was enormous fun. In those wonderful years of 1937 and 1938 everything seemed to go our way. The weather was kind and I look back on the whole period, like that at Orfordness in 1935, as one of the happiest and most productive in my life.

Another pleasant responsibility was to give a series of demonstration flights to a number of VIPs. Our work was becoming well known to senior officers in the Air Force and there were frequent calls for demonstrations. Depending on their interests, the demonstrations were made either against ground targets and ships or against aircraft. During 1938 there were literally dozens of such demonstrations and the list of those who flew with us is impressive. We gave flights to all six Commanders-in-Chief: those of Bomber Command, Coastal Command and Fighter Command, and, when they visited Britain, the Commanders-in-Chief of the Near East, Middle East and the Far East. On 4 July 1938 we gave a demonstration of ship detection to Air Vice-Marshal Sir Philip Joubert de la Ferte. This was again in the Anson K6260 and on this occasion Watson Watt accompanied us—it was his first flight with an airborne radar. As usual, they were most appreciative of what they saw and had many helpful suggestions to make. I distinctly remember that after the flight we indulged in a little speculation about what the future might bring. I ventured the opinion that, in view of its potential, one of these radars would soon be installed in the nose of every aircraft. This was 1938 and I can only say that I obtained a very old-fashioned response from Sir Philip and from Watson Watt. They clearly did not believe what I was saying. I have to admit, I half disbelieved it myself.

We were seeing much less of Watson Watt than in earlier years. He had moved up to become Director of Communications Development at the Air Ministry. We sadly missed his sage advice and the detailed discussion of technical matters which had been so characteristic of his time at Bawdsey. Watson Watt's move to Headquarters brought an important compensation. As ever, he was the consummate salesman and the tireless advocate of radar in all its forms. He was now in a position to do this right at the top, where it mattered most. A P Rowe replaced him as the new Superintendent. He was the complete Civil Servant who knew how to work the Air Ministry machine and keep the filing system in order. Unfortunately, his speciality was bomb ballistics and his knowledge of electronics was minimal. When it came to a discussion of the accuracy of a range-finding system or the best way to go in radio direction-finding, he was completely lost. There is no doubt that he was highly motivated about air defence in an abstract sense, but the details were beyond him. If the truth be told, Rowe never came to terms with people like Wilkins and myself who had done the original work at Orfordness and Bawdsey. Unlike Watson Watt, he seemed to resent any views we might express about the technology of air defence and never, in subsequent years, did he reach any kind of rapport with the pioneers.

We made slow, but constant, improvements in the performance of our airborne radars. During 1938, the principal instrumental change was the substitution of Western Electric 4304 valves for the 'giant acorns' in the transmitter. These gave us an increase in peak power to 1 or 2 kilowatts, with a corresponding increase in the maximum range of detection on ships, which now became a respectable 12 or 15 miles. We were always on the look-out for new short-wave transmitting and receiving tubes with improved performance and these nearly always came from the United States of America, usually from Western Electric or RCA. Although we wished it were otherwise, it was only later, sometimes very much later, that similar valves became available from British sources. With the improvement in sensitivity, we could now just detect the Cork Lightship off Felixstowe from the roof of the White Tower at Bawdsey; it was a tiny target at a range of just over four miles. This was enough to stimulate the Army contingent at Bawdsey to begin construction of a Coast Defence radar—the so-called CD set which was later coupled to the Coast Artillery Batteries. Bill Eastwood left us to pursue that project under the general guidance of Butement and Pollard of the Army Radar Group which had been set up a short time before. Airborne radar had spawned its first offspring. This was a process which continued in later years when the CD radar led to the so-called CHL set for the detection of low flying aircraft, which in turn led to many other radars, all operating on the 200 Mc/s band.

For use by Coastal Command, ASV evolved in two main directions. The first was the homing mode, in which the radar was aimed in a forward direction and left–right indications given by means of overlapping beams as in figure 3.2(*a*). It was some time before we settled on the best configuration of aerials but the general objective was clear by the middle of 1938. The other configuration was the search mode, in which narrow beams were radiated sideways from the search aircraft as in figure 3.2(*b*). This had two advantages: the maximum range of detection was increased as a result of narrowing the beam and the aircraft could cover a far greater search area in a single sweep. In practice, an aircraft would use the sideways-looking aerials for search and, when a target was detected on the beam, the aircraft would turn through a right angle and home in on it with its forward-looking aerials.

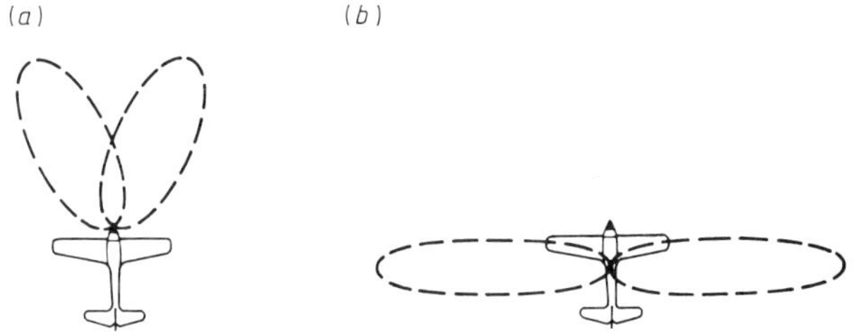

Figure 3.2 (*a*) Forward and (*b*) sideways-looking radar patterns for sea search as provided on aircraft of Coastal Command.

For the first trials, this system was installed only on the port side of the Anson K8758. The transmitting antenna was a six-element Yagi fitted on the wooden tailplane while the transmitter was tucked away in the rear end of the fuselage. The receiving array was a similar Yagi carried on a piece of four by two rigged out of a window amidships. Tests over the sea off Bawdsey gave us maximum ranges of up to 20 miles on medium sized ships and about 30 miles on the coastline. Thus encouraged, Touch and I took the Anson down to the English Channel where we took some historic pictures of our old target, the aircraft carrier *Courageous*, on 9 and 10 May 1938 as it was exercising in mid-Channel. We flew from west to east past the *Courageous* and, using a hastily assembled photographic recorder, produced several pictures, one of which is reproduced in figure 3.3. This is the only copy which now exists on file. It is unfortunate that the original cannot be found because it was much more impressive. In the distance behind the *Courageous* it showed a clear map of the Isle of Wight, the Solent and the surrounding coastline. This was the first ever sideways-looking radar picture and it demonstrated, if only crudely, its potential as a mapping device. Had we known about aperture synthesis in those days, we could have made a dramatic improvement in picture resolution, but this was a technology which took another 20 years to evolve.

The Fleet Air Arm had also become interested in ASV and we made several trips to Lee-on-Solent to demonstrate the equipment to them

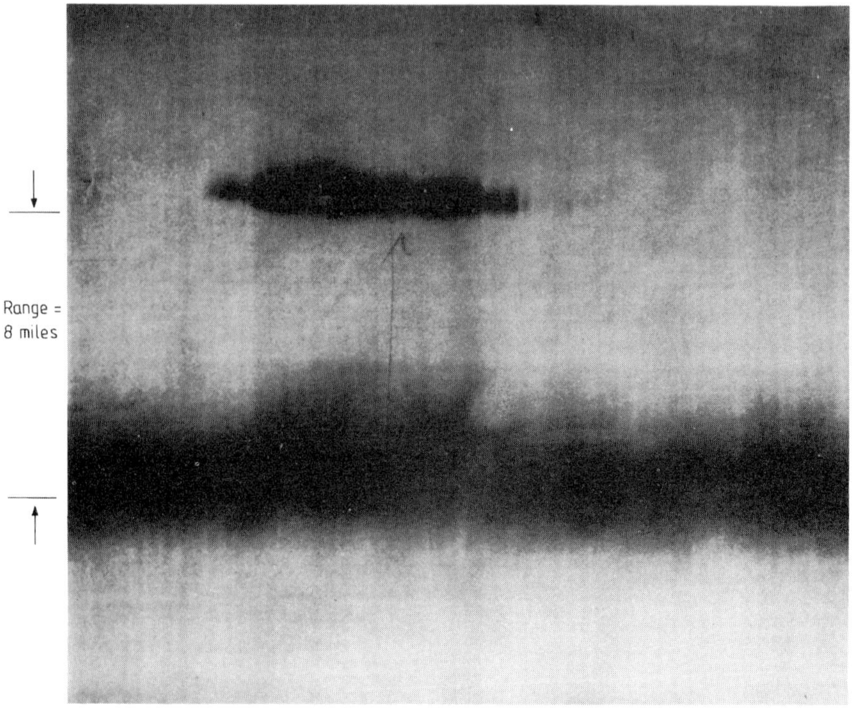

Figure 3.3 The first sideways-looking radar picture taken in the Anson K8758 on 9 May 1938. It shows a saturation signal from the aircraft carrier *Courageous* at a range of eight miles.

using shipping in the Channel. The officers principally concerned were Commanders Couchman, Cecil and York and they made a series of flights with us. The question of what kind of display they needed was much under discussion. Unlike the Air Force, which was attracted to the simple left–right homing system of figure 3.2 (*a*), they were quite definite that for Naval purposes the requirement was for an all-round system which could keep track of a ship behind the aircraft as well as in front of it. The Plan Position Indicator, or PPI, which would give a plan of all the ships around an aircraft, was still only a gleam in the eye, but we tried to simulate it with a rotating dipole system in which direction was determined by the two minima in the polar diagram, as in figure 3.4.

We put a great deal of effort into this device and our two Ansons sprouted an astonishing variety of antennae rotating in the slipstream. The project bristled with difficulties and we never achieved a fully satisfactory rotating display. The most practicable device we arrived at was a range–azimuth display similar to figure 3.5; this later

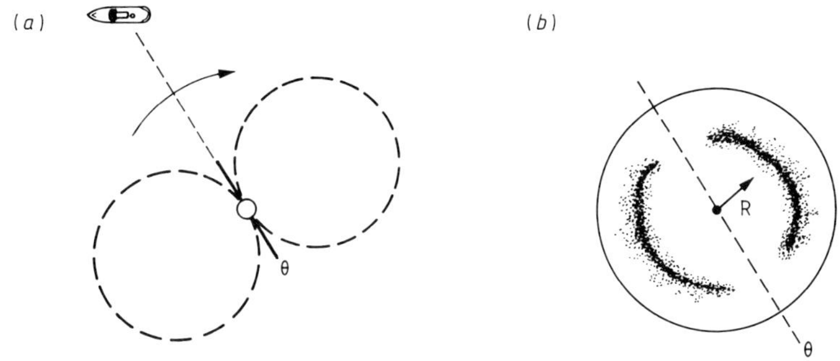

Figure 3.4 An attempt to provide all-round ship detection for the Fleet Air Arm in 1938. (a) The antenna pattern from a rotating dipole. (*b*) An experimental rotating display in which the minimum signal indicates the direction of a target.

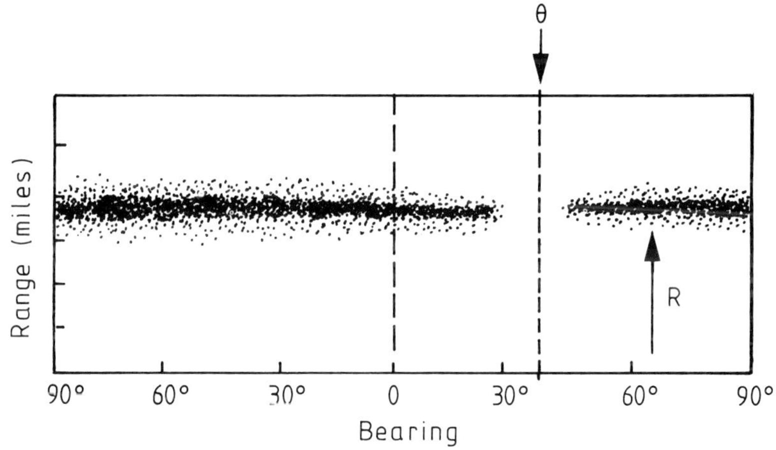

Figure 3.5 The experimental range–azimuth scan ultimately used in the Anson K8758. This later became known as the B-scan and was much used on centimetre-wave radars.

became known as the B-scan. In this display, range was measured along the *y*-axis and bearing along the *x*-axis, the actual bearing being shown by a minimum in the pattern rather than by a maximum. We flew this system in the Anson K8758 for some time but eventually found that, due to losses in the rotating antenna, the penalty in performance was too great. The maximum range of detection on a typical ship was about half that in a forward-looking system, so, reluctantly, we abandoned it in favour of the overlapping beams of figure 3.2 (*a*).

Tales of 'D' Flight

I have already described in Chapter 2 how the allocation of our first two Ansons was the foundation on which the whole of airborne radar was built. The original arrangement for the detachment of these aircraft from 220 Squadron at Bircham Newton remained in force until well into 1938. Pilot Officer Smith and the four Sergeant Pilots, Smith, Naish, Argent and Newman, were all highly competent pilots who served us extremely well. They remained with us until well after the outbreak of war and there is no doubt that the success of the early radar days owes a tremendous amount to the ability and the dedication of these five. Pilot Officer Smith, a young graduate from the RAF Training College at Cranwell, hardly 20 years of age when he joined us, was an especially likeable character. He was tall and slim with a freckled face and gingery hair which quickly earned him the nick-name 'Blood-orange' Smith. He had all the bravado and the ability of the new generation of officers who came into the Air Force about that time. We discovered later that he was also of interesting lineage. His father was an accomplished artist, Sir Matthew Smith, who was interned in France during the war. When he returned to Britain in 1945 he became a distinguished and highly respected President of the Royal Academy.

Air Force officers at that time were encouraged to take up sailing as a hobby and many Martlesham pilots indulged in the sport at the nearby village of Waldringfield—a delightful spot a short distance from Martlesham Heath on the River Deben. Smithy was one of the keenest of these, owning a 12 foot National dinghy in which we spent many a happy afternoon. During the height of the season, Smith would put his dinghy on a trailer on a Friday afternoon, hitch it to his Ford 10 and take off for regattas, either in his native Norfolk or as far afield as Chichester and places like Falmouth in the far south-west. More often than not, he would return on the Monday morning with a trophy or a large bowl saying 'Look what I won at the Falmouth regatta.' If, in what follows, I occasionally overstate the ability of our pilots, it is probably due to nostalgia for the wonderful times we had together. With such an effervescent group, anything could happen and, as we shall see, there were bound to be a few hitches from time to time.

In the early stages of our air-to-air tests it was our practice to have the target, usually the K8758, patrolling between two fixed points on the coast at 15000 feet. The intercepting aircraft, the K6260, would do a series of interception runs on the K8758 until the purpose of the flight had been accomplished, or until we became exhausted. Until well after the war had started, no oxygen was available to the crew on such flights; we were not even told that oxygen was necessary. I can recall more than one occasion on which we took the Anson to its ceiling, which was about 19000 feet. It took us 20 minutes to reach 15000 feet and another 20 minutes to do the last 4000 feet. Many were the splitting headaches and the occasional case of disorientation we suffered as a result of this, but in those days no one was much concerned, least of all ourselves. I do not know what the medics would say about it these days, but there is a message in there somewhere. During such trials, the crew of the intercepting aircraft were fully occupied, but the crew of the target became a little bored, especially if the exercise went on too long. It became the practice when the exercise was over to have an unofficial competition between the two crews. When they got within gliding distance of Martlesham, they would cut their engines back to tick-over and the crew which arrived at the field and landed nearest the perimeter without touching the throttle was the winner—the prize a whole half-crown. This was probably contrary to regulations, but after 50 years I doubt whether anyone is going to bring any charges.

A characteristic of the Ansons was that the undercarriage was hand-operated and there was no warning hooter to indicate whether the wheels were up or down. On landing, either the pilot or a crew member had to make 120 turns of a hand crank alongside the pilot's seat to bring the wheels down. With such an arrangement something was bound to go wrong sooner or later. On one of our practice flights, Smithy in the K8758 was just clearing the boundary, well satisfied that he had won the coveted half-crown. There was an ominous thump and the sound of tearing wood and metal; in the excitement of the moment, he had forgotten to put the undercarriage down! The Commanding Officer of Martlesham at that time was notoriously bad-tempered and had seen the whole episode from the far end of the aerodrome. He came tearing across the field in his staff car and drew up alongside the stricken Anson. Smith crept out in sheepish fashion, the dust still settling around him. The CO came up, held out his hand and said: 'Congratulations, Smith. An excellent landing under difficult circumstances.' He thought the undercarriage had stuck. Smith was not to get away with it entirely scot-free. The aircraft was not a Martlesham machine; it was still on detachment from Bircham Newton, so he had to return to his home base a few weeks later to attend a court of inquiry. He returned a rather chastened man.

We did not hear the sequel to that story until many months later when we gave a demonstration of ship detection to none other than Air Chief-Marshal Sir 'Ginger' Bowhill, the Commander-in-Chief of Coastal Command. Smithy insisted on doing the flying and the flight went off without a hitch. We gave our standard demonstration with several runs on the Butter Boat and the Commander-in-Chief was duly impressed. Smith made a meticulous landing back at Martlesham and the Commander-in-Chief was generous in his praise of the operation. After he had taken his leave, I turned to Smithy and said what a fine man he was. 'Yes' said Smithy 'a marvellous fellow. He was the one who tore me off a strip for landing two Ansons with their wheels up.' *Two* Ansons? What we did not know was that when Smith went back to the Court of Inquiry at Bircham Newton, he had borrowed a Squadron aircraft to keep his hand in and had landed that one with its wheels up too!

That was not the only incident involving the K8758. After an exercise late one afternoon, this aircraft, with two of our Sergeant Pilots on board, had gone off on a sortie of its own, probably to do some low-flying practice over the sea, which as Coastal Command pilots they were encouraged to do. They did not return until after dark and, in the absence of landing lights, they overshot the runway at Martlesham, ran over the main road which went through the camp and sheared six feet off one wing tip on the stay wire of a telegraph pole! The Anson had a metal fuselage and a wooden wing, which under such circumstances sheared off like matchwood. It was relatively easy to repair and a new wing tip was spliced on within a matter of weeks but I must confess that I was never entirely comfortable flying in that aircraft again. The incident also earned me a sharp reprimand from higher authority, probably richly deserved. Someone up there had noticed that while we were piling up an impressive number of flying hours, we made a habit of sneaking into Martlesham just before darkness fell and we were told in no uncertain terms to cut it out.

We had another nagging problem, and this was with ignition noise. The success of an airborne radar depends entirely on the complete suppression of ignition noise from the aircraft engines, which after all are within a few feet of the receiver. We seldom had trouble with the Anson K6260 but, for some reason, the K8758 was always developing excessive noise. When this happened, any developmental work being done on that aircraft was stopped dead in its tracks. We had friendly relations with the staff of the Electrical Department at Farnborough who were expert at fixing the fault. When the ignition noise on the K8758 became too great, we would ring up Farnborough, tell them we had a problem and ask if we could come down and have it fixed. They were invariably most accommodating. It was only a 30 or 40 minute flight to Farnborough and the next day we would be off at the crack of dawn. By mid-morning, Farnborough would have fixed the harness and we would be back at Martlesham before lunch. We would be in business again that afternoon and would continue our work until near nightfall.

The new Superintendent at Bawdsey came to hear about these visits and they did not appeal to his tidy mind. He dug up an age-old Air Ministry order which said that an RAF aircraft should not land at an airfield, other than its own base, without giving three days' notice in writing of its intention to do so. He sent us a memorandum saying that what we were doing was totally unacceptable and would we please abide by Air Ministry orders. I am afraid we put his memorandum in the waste-paper basket where it belonged and continued to do things our own way.

Through most of 1938 the Ansons were employed largely on work against surface ships. However, as the threat of war came closer, the air interception work became more and more urgent. The need arose for an aircraft with a higher performance than the Anson and one which bore a closer resemblance to a fighter. As I have explained, even as late as 1938 there were no firm views at the Air Ministry or at Fighter Command as to what a night fighter should look like and we were told that the Fairey Battle was as good a choice as any. It had the new Rolls Royce Merlin engine which also powered the Spitfire and Hurricane. It had ample room for equipment and an observer and, at a pinch, two observers could be packed into the rear cockpit. Although not fast enough, it was a good deal faster than the Ansons, which were delightful aircraft but rapidly becoming obsolete.

We therefore put in for a Battle and, as had previously happened with the Ansons, we were promptly rewarded with two. These were the K9207 and K9208 which were to serve us well right up to the start of the war. We also gained two additional pilots with experience of aircraft of that particular type. They were Flying Officer Rayment and Sergeant Wright. According to Air Force custom, Sergeant Wright could only be called 'Wilbur' and that was his name from then on. Like our earlier arrivals, our two new pilots were much younger than the average at Martlesham, but they were exceedingly competent and served us well. Rayment, as we shall see, had a flamboyant touch about him, but 'Wilbur' Wright was as solid and reliable as his esteemed namesake of 30 years before. Our other pilots rapidly became competent in handling the Battles, but Rayment and Wright remained the principal exponents. With the arrival of the Battles, it was no longer reasonable for the radar flight to be a detachment of 220 Squadron and it was reconstituted as a regular Flight of Martlesham Heath with 'Blood-orange' Smith, still a very young Flying Officer, as the first Flight Commander. It was the practice at that time to name the different flights at Martlesham alphabetically. The Performance Testing Flights were respectively 'A Per T', 'B Per T' and 'C Per T', so the radar flight naturally fell next in line and became 'D Per T'.

As recounted in the next chapter, 1939 was rapidly approaching when we became heavily involved in the final development of air-interception radar. On scientific and technical matters, Touch had long been my deputy and looked after things on my frequent absences from Bawdsey. He was also principally responsible for receivers and indicators, ably assisted by Hanbury Brown. Hibberd was in charge of transmitter design until he, in turn, departed to take charge of the production of airborne radars at the Air Ministry. Although I always seemed to be drawing new organisation charts with pretentious titles on them to meet the demands of the Superintendent, in fact the organisation of the airborne group remained in the free and easy style of Watson Watt and the staff was switched from one project to another as the degree of urgency demanded. As 1938 drew to a close, Hanbury Brown became increasingly drawn into flying the air-interception equipment with Keith Wood and myself.

When the radar flight became 'D Per T', and therefore a full section of Martlesham Heath, 'Blood-orange' Smith was considered too young to be in charge of such an important flight and a more senior officer, Squadron Leader Butler, was appointed. He had a wealth of flying experience, much of it overseas, but as frequently happens in the Air Force, this was only a temporary arrangement and he was soon replaced by Squadron Leader Hyde—a permanent member of the Martlesham staff with a long and distinguished record of test flying. He proved to be a jovial, well respected and extremely competent member of the band and remained in charge until well after the start of the war. He went under the nick-name 'Hetty'—the origins of which are lost in history—but it had nothing to do with a lack of virility. He was very much a married man with several charming daughters. He remained in charge of the flight through some exceptionally difficult times and was successful in holding it together and maintaining the morale of the troops during subsequent moves to Perth, St Athan and Christchurch.

Hyde finally left the flight in 1940 or 1941, at his own request, to undertake active service. He became the Commanding Officer of the first squadron of Manchester aircraft to become operational. Like several other aircraft introduced about that time, they were not tried out as well as they might have been and he was shot down and became a prisoner of war in a German prison camp. He did not take kindly to confinement and quickly made an escape, only to be recaptured shortly afterwards. He soon became known as a confirmed escapee and was moved to successively more secure camps, finally ending up somewhere in East Germany. He escaped three or four times before finally getting away, riding the axles of a freight train to a Baltic port from which he ultimately made his way by ship to Sweden and thence back to freedom in Britain.

Meanwhile, let us return to 'D' flight. About this time, Smith tired of

Figure 3.6 The personnel of 'D Per T' flight at Martlesham Heath in late 1938. It shows the Commanding Officer, Squadron Leader Butler, in the centre; to his left, Flying Officer Smith, Hanbury Brown, Flight Sergeant Turner, then Sergeant Pilots Naish, Argent and Smith. The other two civilians shown are Mills on the far left and Keith Wood, far right, both members of the Bawdsey airborne team.

sailing, sold his boat and, jointly with his elder brother, invested in a light aircraft—a Klemm Swallow, with a Pobjoy engine of minimal proportions. With a wide range of Air Force aircraft to choose from, they decided on yet another one for their own amusement. They parked it in a corner of a hangar at Martlesham and took it on jaunts around the countryside at the weekends. They caused great amusement at Martlesham by taking it up to a safe height over the field and staging a competition. One of them would put it into an impossible attitude—upside-down or going straight up—at which point he would cut the throttle, abandon the controls and leave the other brother to sort it out. Why they did not break the aircraft up, we could never understand.

Soon afterwards, Smith's brother was posted to Aden and he decided to fly himself there in the Klemm. With one of the Smiths at the controls it was bound to be an epic flight. He went down to Lymnpe—the acknowledged field for the start of overseas flights—and made his preparations. He took on extra fuel and departed on a day of unusually bad weather. It was extremely rough and soon after take-off he ran into low cloud. He stuck it out for an hour or two and then decided he had better go down and find out where he was. He should have been well into France but when he broke through cloud he was still over water. 'Goodness' he said to himself 'that must have been a terrible head wind.' The only prudent thing to do was to turn around and head for home. He did this and soon came over some green fields. He did not have a clue where he was, so he decided to put down and ask. He found a large field with a pleasant cottage in the corner. After landing, he wandered over to the cottage hoping for a cup of tea and some directions. He knocked on the door and was greeted by a kindly lady—who could not speak a word of English! For a moment he was nonplussed, and then the penny dropped; he was not in England but on the south coast of France and the sea he had been over was the Mediterranean.

It turned out that, entirely by chance, the day he left Lymnpe was 10 February 1938, a very special day that the Air Force had been waiting for. For several weeks they had kept a careful watch on the weather, waiting for a screaming northerly gale. When it came, a Hurricane, piloted by Squadron Leader Gillan, was flown from Edinburgh to London at an average speed close to 450 miles an hour. There was something like a 120 mile an hour tail wind and it gave the RAF some welcome publicity for their newest fighter. With that kind of wind behind him, Smith had made exceptionally good progress.

His next stop was Malta, which he reached without incident, followed by a long leg into North Africa. The inevitable happened; somewhere over the desert his engine stopped and he put down in the sand. It happened that this was when Italy was being very belligerent in North Africa and making tracks for Abyssinia. In a very short time, Smith was picked up by Italian troops who took a very poor view of a Royal Air Force officer flying over their territory in uniform. They impounded his aircraft and locked him up while they decided what to do with him. About a week later, General Balbo heard about the incident and asked for him to be brought to his Headquarters in Tunisia. He listened to his tale very sympathetically and then fed him a superlative dinner. He told the troops to fill up his aircraft and sent him on his way to Aden with *carte blanche* to land and refuel at any Italian base along the way. Some years later, Mussolini saw fit to join forces with Adolf Hitler, with disastrous consequences. If the decision had been left to his Generals, one wonders whether they would have decided the same way.

Returning to the fortunes of 'D' flight, the number of aircraft grew with the addition of a Harrow to act as a more realistic target than the Ansons or Battles we normally used. This was followed by another Battle for general duties and a Magister for communications purposes. Gradually 'D' flight became known as one of the busiest flights at Martlesham Heath, but we were not without our lighter moments. I have mentioned earlier how, during interception trials, the exercise would end with both the target and the interception aircraft at 15000 feet somewhere along the coast not far from Bawdsey. This work was now being done by Battles with the exuberant Flying Officer Rayment at the controls. At the end of an exercise he was liable to say 'Let's go down and take a look at Rowe in his office.' Rowe at that time had an office in Bawdsey Manor overlooking the sea, perhaps 50 feet above sea level. Rayment would put the Battle into a vertical dive from 15000 feet and scream down at well over 300 miles an hour. We would level out about 50 feet up, near the mouth of the Deben, and roar past Rowe's window less than 100 feet away. The noise must have been deafening. Rowe was subsequently heard to complain about those Martlesham pilots who, by such inconsiderate behaviour, disturbed the work of an important research establishment. I doubt whether he ever discovered who was responsible but he may have had a sneaking suspicion.

Introduction of Alternators as the Primary Power Source

Towards the end of 1938 we could foresee that AI and ASV would soon come into service in Fighter and Coastal Command aircraft and the problem of how to supply them with electrical power became an important consideration. The standard RAF generator at that time was a 500 watt, 24 volt DC machine and no other type was contemplated. Most of its output was taken up by existing aircraft services and there was no power to spare to run a radar set, however economical. The radar equipment was going to require such a variety of voltages that an alternating supply was essential. A way out of the dilemma existed on two-engine aircraft where it was customary to fit only one DC generator, leaving a spare drive-shaft on the other engine. Adding a second DC generator and using this to drive an alternator was one possibility, but that would be less than 50 per cent efficient and we needed all of the 500 watts. So there was no option but to go to an engine-driven alternator. The problem was how to get one into service.

An Air Ministry Committee had been sitting on the question of alternating current supplies for years without getting anywhere. They regarded it as an either/or problem—either one used a DC system or an AC system. The forces in favour of one or the other were so evenly balanced that there was a deadlock and no effective action was possible. I discussed the problem with Watson Watt who advocated the direct approach to a manufacturer. I made enquiries which indicated that Metropolitan-Vickers (Metro-Vickers) in Sheffield could probably make the alternator we required. I made an appointment to see the Managing Director, Mr Fletcher. As time was short, I decided to fly up to Sheffield in one of our Battles towards the end of October 1938. Sergeant Wright was the pilot and we landed at the nearby airfield of Finningley. It happened that we did not have a spare DC generator back at Martlesham, so after landing I took a spanner, whipped the generator off the Battle and took it hot-foot to the Metro-Vickers Sheffield Works. I laid it on the Managing Director's table and outlined what we wanted—an alternator, not a DC generator, looking exactly like the one on the table, occupying the same space and using the same fixing bolts and spline shaft. We could accept almost any voltage within reason, but 100 volts was a good figure to shoot for. We knew that with changing engine revs we would have to tolerate something like a two-to-one variation in frequency, but this was no problem. Also, for economy in transformer iron, a high, rather than a low, frequency was called for, say 500 or 1000 cycles. Needless to say, as engine revs changed the voltage had to remain more or less constant. Lastly, our total power requirement was certainly not less than 500 watts but anything over that figure would be very welcome. What could be done?

He called in Tustin, his Chief Engineer (who later became Professor of Electrical Engineering at Imperial College), and put the problem to him. Tustin went away to do some calculations. After an excellent lunch in the Managing Director's dining room we gathered again and Tustin was very confident about meeting the requirements. He said that the preferred voltage was 80, the frequency range would be 1200 to 2400 cycles and he could be reasonably certain the machine would deliver about 800 watts! The weight would be about the same as the DC machine and he saw no problem in matching the fixing bolts and the spline shaft. Nothing could be better and we settled for that specification on the spot. These were the figures which were used throughout the war and did not change until a long time after the war ended. Production of the machines presented few problems and Metro-Vickers thought that, if pressed, they could produce the first models in one month. I promised to get the appropriate papers to them in a matter of days.

I returned to Finningley with the generator under my arm. By this time it was getting late in the day and we did not have time to reinstall the generator in the Battle. We started up on a battery cart and decided we could do the return trip to Martlesham on the aircraft accumulator. The Station Commander was there to see us off and he expressed concern about the failing light. Dusk was already beginning to fall. We said not to worry but asked him to telephone Martlesham Heath to say that it might be dark when we arrived. There would be no landing lights and no moon, but we would feel our way in. We took off for home, not exactly singing and dancing but reasonably confident we would make it. As we proceeded south, the cloud base got lower and lower and it was this which eventually was our undoing. Navigation was difficult in the poor light and we soon realised we were south of our intended track. We were down to about 800 feet in the Fen country north of Newmarket when quite suddenly we flew through a huge flock of wild geese and hit one of them with a resounding thump. It must have been a large bird because it swung the aircraft appreciably and made a considerable dent in the leading edge about six feet in from the wing tip. It was nearly dark and this shook us up a little. Sergeant Wright decided to make an immediate landing. There was flat, well ploughed land beneath, so he picked a good long paddock and put down. He made an excellent landing and we were very content to have our feet back on the ground. There was a small village nearby and the accepted drill in a case like that was to telephone the nearest airfield for assistance. Sergeant Wright and I pooled our money to pay for a telephone call. Our combined financial resources reached a total of one shilling and fourpence halfpenny. I stayed to guard the aircraft and Sergeant Wright headed for the village clutching the money in his hand.

The village turned out to be Barrow, just south of the Newmarket–Bury St Edmunds road, and the nearest RAF station was Mildenhall, where the duty officer reacted promptly and sent a guard and transport. It transpired that we must have flown right over Mildenhall a few minutes before landing and we were extremely unlucky not to have seen it in the failing light. In no time at all, the aircraft was surrounded by sightseers from the village; some kindly souls were good enough to bring piping hot tea and buns and even blankets, which were very welcome in the chill night air. Then came the Air Force guard and transport back to Mildenhall, where we put up for the night.

The plan was that Sergeant Wright would fly the Battle out in the morning, while I would head for London to arrange for orders to be placed for the alternators. The station adjutant gave me a voucher for the train journey and in due course I arrived at King's Cross. From King's Cross to the Air Ministry in Kingsway was a journey of several miles and I did not have a penny to my name. It was a very pleasant morning and I walked. I reported to Watson Watt, told him the good news from Metro-Vickers and orders for eighteen engine-driven alternators were placed immediately. I then borrowed £5 from the great man and made my way happily back to Bawdsey.

A model of the alternator was running successfully on the bench within a few weeks and the first deliveries were made within a month.

They more than met our expectations and we put them into immediate use. We never had to worry about power problems again. The order for the first eighteen alternators was completed before the outbreak of war and a repeat order for 400 was made in October 1939. Without this machine, airborne radar simply would not have been possible, or at best it would have been severely handicapped. The alternator was made in hundreds of thousands during the war—the total British production reaching a figure of 133800 units. The alternator was adopted by all the Allied Air Forces, including the Americans. They finished up using two standards, a 400 cycle alternator to drive US-produced equipment and the British alternator on the many occasions on which they carried British airborne gear.

4

Air-Interception Trials in Battles and Blenheims

At the beginning of 1939, the maximum range performance of air-to-air radar was already two to three miles and getting close to the required figure of three to four miles. The most urgent tasks, therefore, were to add directional information, in azimuth and in elevation, and to ensure that the minimum range performance was adequate, namely, less than 1000 feet. The controls of a typical radar set were very similar to a modern television receiver, with knobs for tuning, focus and brightness. It was generally agreed that a radar operator would be able to handle these without difficulty, but it was desirable that bearing and elevation be given to him directly on the screen, not by the manipulation of additional knobs as was still the practice in conventional ground radar systems. These were pioneer days and there was no prior experience to indicate just what an operator could and could not do at night in the back of an uncomfortable and extremely noisy aircraft. There was no doubt that whatever system was adopted, it had to be a simple and obvious one.

At the beginning of 1939, the possibilities for indicating elevation and bearing of a target aircraft were three in number.

(1) The first was a range–azimuth display system similar to that shown in figure 3.5. Unfortunately the penalty in maximum range was too great, so this otherwise attractive scheme was abandoned.

(2) We put a great deal of thought into a phase-comparison method in which the direction of the target was determined by the phase of the signals received on two spaced antennas. Unfortunately, variable phase changers simply did not exist on 200 Mc/s and, with our small staff, we simply could not do the necessary development work. This is where the skills of a Bainbridge-Bell might have been invaluable.

(3) The next possibility, and the one finally adopted, was an overlapping beam system, of the kind well known since the days of the Lorentz landing system. This was an arrangement in which a pair of overlapping beams was generated in azimuth and another pair in elevation, as in figure 4.1.

The most natural form of display to go with this was that shown in figure 4.2 (*a*) and (*b*). In this display, an aircraft echo appears in the

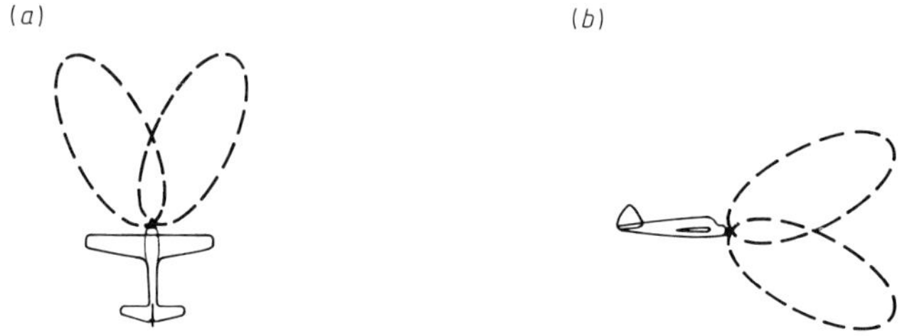

Figure 4.1 The overlapping antenna patterns in (*a*) azimuth and in (*b*) elevation of 200 Mc/s air-interception radars Mark I to IV.

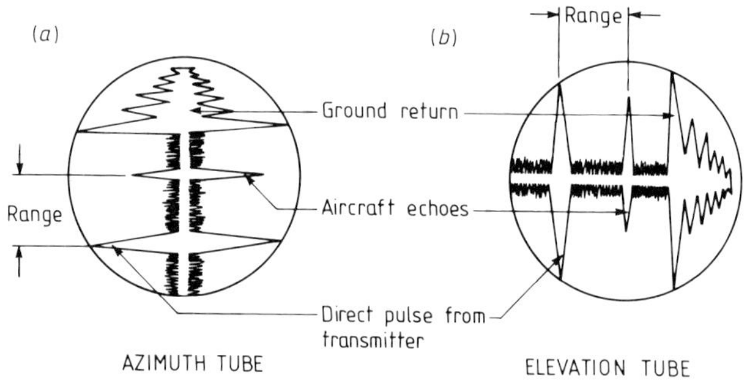

Figure 4.2 (*a*) and (*b*) The 200 Mc/s air-interception display patterns in azimuth and elevation respectively. In (*a*), the target aircraft is shown to starboard and in (*b*) it is above the fighter aircraft.

space between the direct pulse from the transmitter and the large signal at the end of the trace. This was the signal which came from the ground below and ahead of the aircraft—the so-called 'Xmas tree'. It is clear from this diagram that when the target aircraft echo runs into the ground return, it will be completely obscured. In other words, the maximum range of detection on another aircraft is limited by the height above ground, irrespective of the sensitivity of the equipment.

At a flying height of 15000 feet, the maximum range would be almost three miles and this would not present too many problems. However, at a height of 10000 feet or less the maximum range was less than two miles and this did not leave the fighter much room for manoeuvre or for correcting small mistakes. This maximum range limitation had not existed in the very first airborne set, the RDF 1.5 system flown in the Heyford over Bawdsey in 1936, but the authorities chose to ignore it. The problem of improving the maximum range performance of airborne radar remained a constantly recurring theme throughout the pre-war period. The solution was clearly to go for increasingly shorter wavelengths, but the means for carrying this out did not become available until 1940.

In the example given in figure 4.2 (*a*), the target aircraft is shown as being to the right of the fighter aircraft and in the elevation display, figure 4.2 (*b*), the target is above. In practice, the radar operator passed this and the range information in a continuous flow to the pilot, who then carried out his interception according to these instructions. The display shown in figure 4.2 was the one decided on in the spring of 1939 and subsequently adopted in all metre wave air-interception radars. It was used by all the Allies and was copied almost exactly by the German Air Force. Switching signals out of the back end of the receiver to achieve this display presented no problem. It was clear that a simple mechanical switch would do the trick. But we shuddered at the thought of a mechanical switch at the front end of our super-sensitive receiver. We were quite sure that switching noise would swamp the system. We therefore dreamed up all kinds of alternatives —usually electronic methods which in practice turned out to be far too complicated or led to excessive losses. In desperation, we finally took the plunge and Touch built a two-position radio-frequency switch which made a good match with our transmission lines. To our complete surprise, it worked first time—and there was no noise problem. If there was a noise pulse on switching over from one aerial to the other, it was non-synchronous and seldom, if ever, appeared on the screen. So, two problems were solved in one stroke. The two-position switch gave us the left–right indications necessary for a forward-looking sea-search radar and a four-pole version gave us both azimuth and elevation for the air-interception application. This was accomplished in May 1939, dangerously close to the outbreak of war; by 1 June 1939, we were flying a complete air-interception system in the Battle K9208 which gave both azimuth and elevation indications, and a corresponding azimuth only system in the Anson K6260 for sea searching.

It had always been recognised that the minimum range performance of an air-to-air radar would be crucial. At that time, the thought of firing blind at a target aircraft at night was simply out of the question. The Air Force made it an absolute requirement for a pilot to identify his target quite positively before opening fire. There would be just as many friendly aircraft in the air as there would be enemy aircraft and a high probability that a night fighter would intercept a friendly aircraft. This was a situation in which guesswork could not be tolerated. It was essential, therefore, that the minimum range should be small enough for the pilot to see and identify his target by eye before opening fire. In our discussions with the Air Ministry it was generally agreed that a minimum range of 1000 feet was the figure to aim for. This figure was based on a series of night tests which had been made at Farnborough several years before; they found that on moonlit nights a target aircraft could sometimes be seen up to 2000 feet away, but on really dark nights or on nights when the moon was obscured by cloud, a range of 1000 feet was required and this was the figure we aimed for.

During our own air-to-air trials in 1938 we knew that we were close to the required figure, averaging around 1200 feet, and we saw no problem in reducing it. There were some rough rules of thumb which spelled out what could be done and without going into detail, at an operating frequency of 100 Mc/s, a 1 microsecond pulse would give a minimum range of 300 metres or almost exactly 1000 feet. At our operating frequency of 200 Mc/s, the corresponding minimum range figure was 500 feet—which was well within the required figure. We started serious work on the minimum range problem in the spring of 1939, at about the same time as we were tackling the azimuth–elevation problem. Again, it was Touch who arrived at the answer. There was no difficulty in reducing the transmitted pulse to about 100 cycles which corresponded to that 500 foot minimum range; the whole problem was in the receiver, which was liable to be overloaded by the shock of an enormously strong pulse coming from a transmitting aerial only a few feet away. Touch found that this could be greatly reduced by quenching the receiver's local oscillator for the duration of the pulse—a very simple and elegant solution. In the fullness of time this message spread to most other designers of radar equipment who incorporated it in their designs. Reference to Bawdsey reports of that era will show that, at least in Britain, it was Touch who made this advance and the airborne group were the first to take advantage of it.

We checked the minimum range performance of the airborne set by flying against a much more demanding target than another aircraft, namely, the sea surface directly below the aircraft. Easily the strongest echo received in an airborne radar is the direct return from the ground, or in this case the sea, immediately below. Touch checked various configurations of pulsewidth and receiver suppression in the Anson K6260 as it flew at different heights above the sea; an accurate measure of height being obtained from the altimeter at the same time. He found that there was no great difficulty in getting down to a minimum of 500 feet. Indeed, by squeezing things a little he could get down to 200 feet, but in doing this there was a danger of reducing the maximum range. In order to be quite sure we did not prejudice our maximum range performance—a terribly important consideration—we compromised on 800 feet as being comfortably within the requirement.

In case we ran into trouble later, we had a fall-back position. This was to use a slightly longer transmitted pulse for detecting a target at maximum range, and then to reduce the pulsewidth for the final approach; however, the need for this expedient did not arise. I have elaborated on the minimum range question because it was the cause of a misunderstanding later, when the airborne group was accused of having paid insufficient attention to this problem. Some of the production sets did not always meet the 1000 foot minimum range requirement, but this was a matter of adjustment, not of basic design.

Demonstration of Air Interception to the Commander-in-Chief of Fighter Command

With the design of our air-interception radar coming into its final form, we embarked on another series of demonstration flights. By far the most important of these was to the Commander-in-Chief of Fighter Command—the legendary Air Marshal Sir Hugh Dowding. He was the man who in 1935 had made the first grant of £10000 for the development of air-warning radar and the man who was destined to win the coming Battle of Britain. We prepared for this flight with a certain amount of apprehension; the Commander-in-Chief had an awesome reputation as a stickler for correct Air Force procedures and we were not noted for going by the book. He had long since earned the nick-name 'Stuffy' in Air Force circles and was universally referred to by that name. The flight took place one morning in mid-June in the Battle K9208, with 'Blood-orange' Smith as pilot and myself as demonstrator. There was very little room for a second passenger in the back seat of a Battle. In fact the only way to accommodate a second person was to put a plank of wood across the seat and sit side-by-side with both heads under the one black cloth shielding the cathode-ray tubes. There was no such thing as being strapped in and no oxygen.

There were a couple of parachutes somewhere on the floor of the aircraft, which was the best that we could provide. The weather was fine and clear and we went straight to 15000 feet where there was another Battle to act as target. We then made several demonstration interceptions, coming in to the target from maximum range, from positions roughly 30 degrees to port or starboard of the dead astern position. The demonstrations went extremely well and the Commander-in-Chief was well satisfied.

Before the flight, the Commander-in-Chief had emphasised that he was particularly interested in the minimum range performance. As we made the last interception, he said: 'When we get to minimum range, tell the pilot to hold that position so that I can see how close we are.' This we did. When we had settled down at minimum range, Stuffy said: 'Now let's take a look.' For the previous 30 or 40 minutes our heads had been under the black cloth shielding the cathode-ray tubes. I whipped the cloth off and Stuffy looked straight ahead. He said: 'Where is it? I can't see it.' I pointed straight up; we were flying almost directly underneath the target. 'My God' said Stuffy 'tell him to move away, we are too close.' We broke away and headed for home. The Commander-in-Chief was clearly pleased with what he had seen.

This tale would not be complete if I failed to record the details of our arrival back at Martlesham. Needless to say that with the Commander-in-Chief on board there was no nonsense about competing with the target aircraft for a half-crown for a dead-stick landing. The target aircraft had in fact been warned to keep well away and leave us in peace. Smith at this time was fairly well acquainted with the Battle, but his landings were not always the best. As we were on the final approach, I could see that he was much too high and was going to overshoot—I regarded myself as a pretty good back seat driver by that time. 'For God's sake, Smithy' I yelled down the intercom 'Go round again, you are going to overshoot.' Smithy came back promptly and said: 'Not bloody likely, not with Stuffy in the back.' What he and I did not realise was that Stuffy was still on the intercom and heard the whole exchange. Smithy tried to slip off some height but did not succeed very well. We put down with a dreadful thump half way across the field and skidded to a halt. It turned out later that we had broken a tail frame and that the tail wheel had collapsed. We limped back to base with awful grinding noises coming from the rear of the aircraft. Before getting out, I made apologies to the Commander-in-Chief, saying that these young pilots were really very good, but they were often required to fly two or three different types of aircraft in a day. Stuffy at that time had a fearful reputation as a disciplinarian and there were many stories about him. He once put a Flight Sergeant on a charge when, during a hangar inspection, Stuffy found a smouldering cigarette in a fire bucket. Smithy climbed out of the aircraft wishing he hadn't been born, but he faced up bravely to the Commander-in-Chief and said: 'Very sorry about that landing, Sir.' Stuffy took a deep breath and said: 'I always say, the most important thing is to land the right way up' and he said no more about it.

We went to lunch at the Mess where the Commander-in-Chief was entertained by the Commanding Officer and some of his senior pilots. After lunch, Stuffy specially asked that we be left alone to discuss the night interception problem further. We must have continued for the best part of two hours, consuming cup after cup of coffee. Stuffy gave me a long dissertation about the problems of night fighting as he saw them. He had obviously thought very deeply about the subject and I only wished such a discussion could have taken place earlier. Perhaps this was not possible until he had actually received a demonstration flight. I simply do not know whether these thoughts had been with him for a long time or whether he was improvising in the light of what he had just seen.

He covered the whole field, and I will try to summarise what he had to say.

(1) He began by pointing out the obvious—that the night problem was different in nearly every way from day fighting, and that both the aircrew and those in command of a night operation would have to think quite differently about it. For example, the night fighter's task, unlike that of the day fighter, would be long-drawn-out and the aircraft would have to have a comparatively high endurance. Several night fighters would have to be on patrol over each part of the coastline, in anticipation of any raids.

(2) He was quite certain that the pilot would be completely occupied by flying the machine and could not hope to handle radar. This meant a two-man aircraft. Another reason for this was that the pilot would have to be dark adapted. Even under the best possible conditions, it was going to be extremely difficult for the pilot to see another aircraft at night; it would be virtually impossible for him to do so after staring at a relatively bright cathode-ray tube for a half-hour or more.

(3) There might also be a navigational problem. In a long drawn-out interception the night fighter might end up far from his base and, as we had already found hunting for ships over the North Sea, it was extremely difficult to keep track of one's position while weaving about in a search pattern. He even expressed the view that the night fighter might have to carry a third crew member to act as navigator. All of this pointed to the night fighter being a multi-seat, two-engined aircraft. This was the first time I had heard the argument for a two-engined, two-man aircraft advanced with such certainty and with such

authority; in all previous discussions, at the Air Ministry and at Fighter Command, for every advocate of a two-engined aircraft as night fighter, there was at least one vehement advocate of the single-seat, single-engined machine.

(4) Dowding next emphasised the importance of the pilot making a definite visual identification of the target before opening fire. He was aware of the appalling effect on morale if night fighters started shooting down our own aircraft by mistake.

(5) Lastly, he touched briefly on the importance of fire power; he explained that this was even more important than in day fighting. Having once opened fire on a target, a night fighter would seldom have another opportunity of attacking the same target. The victim would immediately dive and take violent evasive action, after which it would be extremely difficult to pick him up a second time.

I had never heard such a clear and definite analysis of the fundamentals of night fighting. Although Stuffy did not give a firm instruction during that visit, after a few days we were told that the Air Force was most unlikely to use the Battle as a night fighter, and that the Blenheim was the most probable choice. In mid-July two Blenheims arrived at 'D' flight with urgent instructions that they should be fitted with our AI. These were the K7033 and K7034, as we shall shortly describe.

Demonstration of Air Interception to Churchill and Professor Lindemann

With war rapidly approaching, Winston Churchill was becoming increasingly prominent and was clearly grooming himself for high office. He was keenly interested in air defence matters and insisted on seeing and being briefed on the latest devices. He visited Bawdsey on 20 June 1939 and was given demonstrations of the chain and coast defence radar systems then in existence. On his way back to London, he stopped at Martlesham to see the AI equipment in the Battle K9208. He did not have time to take a flight, so we laid on a ground demonstration. It takes a youngish man to climb into the back cockpit of the Battle, so we asked the carpenters to knock up a wooden stairway, much like that used on a modern airliner. We placed the K9208 on the edge of the field and ran its radar off a battery cart. As a target, we used another Battle which flew tight circuits around the field at a height of 1500 feet. Meanwhile, all other flying was suspended and in front of an admiring group of airmen the great man prepared to climb into the aircraft. He still had the stub of a cigar in the corner of his mouth and, more or less as an afterthought, handed it to the

Commanding Officer who was standing alongside the aircraft. In other circumstances, it was his job to hand out punishment drill to anyone found smoking that close to an aircraft.

The fun was not over. In his younger days Churchill had a reputation for wearing exotic headgear—he would don anything from the head-dress of a Red Indian Chief to a miner's helmet—and this occasion was no different. He doffed his black Homburg, deeming it unsuitable for wearing in the back of an RAF aircraft, grabbed the forage cap off the head of a nearby airman and climbed up the steps. It was a tight squeeze getting his portly frame into the back seat of the K9208 alongside Hanbury Brown who did the demonstrating. He managed it and put his head under the black cloth shielding the cathode-ray tubes. The target aircraft roared around the field and the radar display showed the changes of range and bearing as it did so. It was as good a demonstration of air-to-air radar as was possible from an aircraft sitting on the ground and the great man seemed well satisfied. He climbed out, reclaimed his cigar and returned the forage cap. I like to think that that cap still exists in Great Britain as a cherished memento, surely the only airman's forage cap worn by someone who would shortly be Prime Minister.

To conclude the visit, the Commanding Officer had arranged afternoon tea at the Officers' Mess, where we now gathered with his senior officers. A gorgeous silver tea-set was produced and some luscious goodies. The gathering was fairly informal and, as time was short, some of the officers were served with tea while Churchill was still being introduced around the gathering. He sat down in a comfortable chair, the pilots standing around him. When asked how he would like his tea, he said 'Tea! Tea! Fetch me a brandy—a big one.' An extra large goblet of brandy was quickly produced and he sat there, a very contented figure with the brandy in one hand and a fresh cigar in the other. He was surrounded by the pick of Britain's test pilots, each holding a cup of tea! Churchill consumed his drink with evident relish and, as if on a signal, stood up and gave a 15 minute oration. I cannot remember exactly what was said, but it was completely off the cuff. It was not quite 'We will fight them on the beaches, we will fight them in the streets . . .' but it was very close. It was entirely suited to the occasion and there was not a man present who was not inspired by his words.

Probably as a result of this demonstration, we received a visit from Professor Lindemann (soon to become Lord Cherwell) a few weeks later. He had long been a personal friend of Churchill's and had been acting as his unofficial scientific adviser. He asked for a flight in the Battle and we arranged this on 10 July 1939. He had already been to Bawdsey that morning and, like Churchill, called at Martlesham on his way back to London. It was late in the day when he arrived and he was quite worried about a dinner appointment in London that evening which he did not want to miss. So we rearranged the demonstration flight to take place along a line from Martlesham to London, so that we could drop him off at North Weald at the conclusion of the test. This was an RAF Station on the outskirts of London; his Rolls Royce was sent ahead to meet him there.

For some reason which I have forgotten, our target was a Wellington of 'B Per T' flight, flown by Flight Lieutenant Slee—an old friend of the radar boys. I think he was already scheduled to do an altitude test that day, so it was decided that we might as well use him as a target. It is of some historical interest that this aircraft was the prototype Wellington undergoing its final acceptance tests. On this particular afternoon it was loaded down with bags of lead shot to 50 per cent over its design maximum weight. Later that afternoon, when we were close enough to have a good look at it, I was amazed. The air was relatively smooth, but from close astern that Wellington looked as if it were made of rubber. The wing tips flexed up and down three or four feet and the fuselage snaked about like a giant fish in a fish bowl. It was a fascinating sight and was the best example of aero-elasticity I have ever observed. Ever since, I have been very partial to aircraft which bend and flex their way through turbulence—as distinct from those which are rigid bodies and drive through without bending. I have an instinctive feeling that if anything is going to break in rough air, it is much more likely to happen to a rigid structure, rather than the one which bends and adjusts itself to the prevailing air currents.

Since the Wellington was carrying such a heavy load, it would take a long time to reach 15000feet so Slee took off early. We undertook to find him by radar somewhere between Martlesham and London. We gave him 20 or 30 minutes start and then followed in the Battle. Hetty Hyde was the pilot and his two passengers, a very portly Lindemann and myself, were squeezed onto the wooden plank across the back seat—the same plank which we had used for Stuffy Dowding, the Commander-in-Chief of Fighter Command, and for Winston Churchill only a few weeks before. We made a normal take-off but were hardly off the ground before it became obvious that something was badly wrong. Masses of black smoke started to pour out of the engine and boiling oil came streaming back over the cockpit. Hetty Hyde said: 'Christ' and snapped back his copkpit cover. The thought shot through my mind: could we ride it out or would we have to jump for it? In the rear cockpit were two seat-type parachutes, but they were on the floor and it would have taken Houdini himself to get into them.

Hetty made reassuring noises from the front 'I think we can make it.' He throttled back and as he did so the smoke subsided but the oil kept coming. With his head out of the cockpit, he did a very sedate circuit and, after what seemed an eternity, we made a smooth landing. The engine was still smoking furiously but the worst was clearly over and the fire-engine which drew alongside as we came to rest did not have to unwind the hoses. I have to say that Lindemann—an old Farnborough pilot from World War I—took all this in his stride. There was simply no panic and no visible reaction to speak about.

I climbed out on the wing and scrambled up to the front cockpit. Hetty's face and shoulders were covered in hot engine oil but otherwise he was unshaken. A quick check showed that someone had left the filler cap off the oil tank and practically all the oil had been pumped out; it had streamed out of the cowling and over the exhaust stubs and it was this which had caused all the smoke. It was a miracle the oil did not ignite. Lesser souls would have quit right there, but both Lindemann and Hetty Hyde were prepared to have another go. After a quick mop down and a refill with oil, Hetty restarted the engine. It seemed to be quite healthy and there was no more smoke, so we prepared for another take-off. This time, everything went well and it was not long before we caught up with the Wellington. We finished by giving Lindemann a very effective demonstration of radar interception and at the end of a long and rather eventful afternoon we headed for North Weald as planned. As we were on the final approach, Lindemann's Rolls Royce came through the Station gates to meet him. The timing could not have been better.

We expected that Lindemann would be impressed, at least with the technical part of the demonstration, but it was not so. He seemed to be in his usual mood of finding fault with everything and did not have a single positive suggestion to make. I wrote later '. . . he seemed rather peeved about the whole thing, and generally gave the impression it was a bit of a fraud'. When he got back to London, he gave us a poor report. Of all the distinguished people to whom we demonstrated airborne radar, Lindemann was the only one who was unimpressed. For reasons best known to himself, he continued to deprecate British radar, at least in its defensive role, for most of the war. Strangely enough, only when it was realised that radar might play an important role in the bombing of Germany did he become enthusiastic about its potential. But that is another story.

Air-Interception Trials in the Blenheim Aircraft

Following the Air Ministry decision to adopt the Benheim as a night fighter, we converted from Battles and began an intensive series of flight trials using the Blenheims K7033 and K7034 as night fighter and target respectively. Compared with the Battle, the Blenheim was a spacious aircraft in which it was relatively easy to install the gear. There was also no difficulty with the elevation aerials, but for the first time we ran into a nagging problem with the azimuth aerials. On the Battle we had used stub antennae on either side of the fuselage. This was a clean configuration which gave a very satisfactory azimuth pattern. However, compared with the Battle, the nose of the Blenheim was short and stubby, the skin was riddled with perspex windows and there were two huge radial engines, one on either side. It was difficult to find a location for the azimuth aerials which provided smooth overlapping beams and Wood, Chalky White and Mills spent a great deal of time swinging the aircraft on the compass base checking aerial patterns. They finally achieved a satisfactory pattern, but it was by no means as straightforward as it had been on the smooth profile provided by the Battle.

It was already July 1939 and things were rapidly coming to a head in Europe. Up to that time, little thought had been given to producing airborne radars in any quantity. Towards the end of 1938, anticipating a demand for production of airborne equipment at some future date, Watson Watt had placed a preliminary order for six airborne transmitters and six receivers with Metro-Vickers and A C Cossor Ltd respectively. These were the contractors already working on construction of the chain air-warning system and they were chosen simply because they had already established procedures for operating under the strict security arrangements which were then called for. They were asked to come to Bawdsey, obtain samples of our airborne equipment or a specification of requirements and to manufacture copies.

Our air-interception transmitter now used British equivalents of the original Western Electric 4304s and our best model was built into the leading edge of the rather thick wing of the Battle K9208. We had a spare ready to install in case of trouble and the Vickers representative asked for it as a model. We were reluctant to part with it as it was our only spare and demands on us were getting heavy. So, we gave him a full specification complete with sketches and photographs. However, he persisted and asked for at least something to take back to Manchester which he could run in the laboratory and obtain a pulsed transmission to measure against. In response to his plea, we gave him an old laboratory transmitter built on a wooden frame; we had once used it in the Ansons but it had long since been discarded. It worked tolerably well, but we made it clear that this was not the design to be copied.

Meanwhile, we gave Cossor a complete run-down on the receiver and specified the sensitivity, band width, output parameters, weight and so on. Once again, we could not let them have a model as we still literally had only the one receiver with the original EMI 45 Mc/s chassis as IF amplifier. They went away confident that they could deliver the equivalent and we were presented with their first six models a few months later. They were a complete failure. They did not come within a factor of ten of the required sensitivity and there was no obvious way of improving them. Furthermore, their weight was astronomical—they weighed more than our complete system up to that time. It was with some regret that we sent them back; we could have used several more receivers at that time. We heard no more from Cossor as potential manufacturers of airborne receivers, but I doubt whether they were much concerned; they were already fully loaded with orders for receivers for the main air-warning system and for Army radars. We did not receive any sample transmitters from Metro-Vickers at that time, but there was a later development concerning this particular company.

Something clearly had to be done about receivers. EMI would have been an obvious choice of contractors, but we were under some kind of restraint not to talk to them. Quite by chance in April or May of 1939, I heard some encouraging news from Edward Appleton, my old Professor at King's College and now the Jacksonian Professor of Physics at Cambridge. He told me that the Pye Radio Company, still hoping there would be a television industry in Britain, had set up a production line for 45 Mc/s TRF chassis and had actually made a trial run. I went hot-foot to Cambridge to see B J Edwards, the Technical Director of Pye Radio, and was rewarded with a remarkable sight—he had scores of TRF chassis of just the type we were looking for.

These used a new valve with an octal base which had not yet appeared on the market. It was the EF 50, a valve which was destined to play almost as important a part in the radar war as the magnetron itself. We were told that the EF 50 was manufactured by Mullard and we were assured that it was in good supply. However, when the Germans started to go through the Low Country in May 1940, a terrible thing happened. It was suddenly discovered that Mullard had in fact run into serious difficulty manufacturing the valve base and that, although the valves we were now depending on bore the Mullard label, all of them came from Philips at Eindhoven. Under circumstances which matched the Zeebrugge raid of World War I, a destroyer was dispatched post-hast to The Hague and, virtually under the German guns, something like 25000 complete EF 50s together with no fewer than 250000 valve bases were brought out from Holland.

Using these bases, Mullard was able to commence manufacture of EF 50s in Britain and the situation was saved. It happened that those valve bases from Holland had nine pins, not eight, and from that time onward the EF 50 was assembled on a nine pin base, not the original octal base. I took a few samples of the Pye chassis back to Bawdsey and we quickly verified that here was a product significantly better than our old EMI chassis; it was also smaller and lighter. Touch and his men quickly put a 200 Mc/s mixer in front and we were in business. B J Edwards did likewise in Cambridge and we were very nearly ready to go into limited production on airborne receivers. In addition to their expertise on receivers, the Pye Radio Company had also embarked on the production of a type of cathode-ray tube suitable for airborne radar. I passed all this information on to Watson Watt at the Air Ministry.

It was now the end of July 1939. We were satisfied that the new installation in the Blenhiem was behaving well and we looked forward to orders being placed for a few prototype models of both air-interception and sea-search radar. Up to that time, six was the number contemplated. What actually happened came as a considerable shock. With virtually no prior notice, we were told at the beginning of August that orders were being placed with Metro-Vickers for 30 transmitters and power supplies, and with Pye Radio for 30 receivers and that these not only had to be manufactured but fitted to 30 squadron aircraft in 30 days—literally by 1 September. This was barely a month after our demonstration to Stuffy Dowding, so the decision making had been rapid, but it bore little relation to the resources available to carry it out.

The setting of unrealistic and unattainable targets was something which plagued radar production until the end of the war. Whether this is the right way to proceed with technological warfare is open to grave doubt. Certainly, if the enemy makes an unexpected move or comes up with an unexpected technical development, a quick and intelligent response is called for. But in conducting long-term plans of the kind required in a national air defence system, there is no substitute for proper foresight. Tizard is the outstanding example of a man who worked out what was likely to happen many years before the event and saw to it that the right steps were taken. Unfortunately, Tizard was not in an executive position in the events now being described.

The original orders for six sets from Metro-Vickers and Cossor were not for complete airborne systems, but for transmitters and receivers only. We ourselves had to fit them together, supply the cables, fittings, and the innumerable bits and pieces which make up a complete system. We managed to persuade the Air Ministry to include in the Pye contract most of the switch-gear, power supplies, control panels and the rest of the components which made a complete assembly. But there were no racks, no brackets, and no cable runs—nothing with which to install the completed equipment in an aircraft. Neither were there any shake tests, altitude tests nor acceptance tests, except of the most rudimentary kind. So much for the primitive conditions under which we worked and for the pious hopes of completing all this in 30 days.

We had another worry. The Ansons and Battles we had used so far, although slow and hardly up to their German equivalents, were wonderfully safe aircraft and ideal for experimental work. The Blenheims were a different matter. They did not come from the Bristol Aircraft Company which designed and built them in the first place but from the shadow factories which had just come into being. I have no wish to malign the dedicated and hard-working people who were doing their best for the country well before war was declared, but the Blenheims delivered to us were not as well built as they might have been. The factories were working under the same kind of pressures which were placed on us and the manufacturers were seldom given enough time to get the bugs out of the product. We were getting the first run of aircraft coming off the assembly lines and minor faults were obvious. Rows of rivets were often missing, the marriage of engine and airframe was not the best and after one or two flights one became apprehensive about what other faults might lie hidden under the skin of those aircraft.

Trouble was inevitable. The K7033 was returning to base one afternoon, trundling along in clear weather at 15000 feet over the tranquil Suffolk countryside, suddenly, the port engine gave a tremendous roar, the revs went up to an astronomical figure—and the propeller flew off. It sailed away into the distance, climbing sedately in an upward direction until it went out of sight. The pilot cut the engine and barely saved it from self-destruction. The Blenheim could fly reasonably well on one engine and there was plenty of height so it got back to Martlesham without further problems. The hydraulics were still operating, the wheels and flaps went down and the crew made a safe landing. It was subsequently found that a tooth had broken off the propeller reduction gear and jammed in the gear casing. It had cut its way into the propeller shaft and sheared it off like a parting tool on a high-speed lathe.

A new engine was installed and the K7033 was back in business, but we joked among ourselves that as the K7034 was the next number in sequence, it might soon perform in the same way. Sure enough, a few weeks later when the K7034 was acting as target, it suffered the same fate, except that this time the starboard propeller was lost. On returning to base, the crew found they had no hydraulics and could not get the wheels or the flaps down. They made a comparatively safe belly landing but the aircraft was a write-off. It was replaced by another Blenheim numbered K7044. The numbers were disconcertingly close.

In less than 24 hours there was an amusing sequel to this last episode. The propeller from the K7033 was never recovered. It must be lying at the bottom of a Suffolk creek or stream to this day. But the day after the K7034 incident, an irate farmer showed up at Martlesham Heath with a very bent propeller in the back of his truck. He told a highly coloured story of how he had heard a whirring noise in the sky. He looked up and saw a propeller coming down rotating at high speed. It then chased him around his home paddock and very nearly decapitated him. Next, it ploughed its way through both his chicken pens and his pigsties, where it caused terrible slaughter. He produced several carcasses to lend weight to his story. I do not know the end of the tale, but the Air Ministry employs legal wallahs to assess how much compensation is paid in such cases; I feel sure that a proper balance was finally struck between how much of it was fact and how much fiction.

Before any of the production sets arrived, Blenheims started to be delivered to 'D' flight, flown in by ferry pilots from factories in the Midlands. When they arrived we had another shock. They were not even the same model as the Blenheims previously sent to us—they were the so-called long-nose version. This meant that many of the fittings and cables we had prepared had to be modified. What was to prove even more of a problem was that the whole azimuth aerial layout had to be changed. This meant more uncertainty and several more weeks' work by Chalky White and Mills, measuring polar diagrams on the compass pad.

Meanwhile, the magnitude of the task we had been presented with bore down on us. If my count is correct, the total staff of the airborne group was just 23 people, including my faithful typist, Betty Holding, and the cleaning man. There were six or eight scientific staff, a corresponding number of technical and assistant staff and positively no one qualified to do aircraft fitting. There was no lack of people at Bawdsey by this time; the total staff was several hundred people, but in spite of the pressure being placed on the airborne group, the number of skilled hands available for airborne work was ludicrously small.

Meanwhile, transmitters and receivers started to arrive from Metro-Vickers and Pye respectively. To our dismay, there was a slip-up at Metro-Vickers. Instead of copying the 1939 transmitter which we had specified, they had copied the 1937 model—the wooden version which their engineer had taken away with him several months earlier to act as a laboratory test piece. When we protested, we were told that during a visit to their Manchester factory Watson Watt had read them his lecture on putting up with third best and insisted that they produce the model which they had working on the laboratory bench, not the one we had specified. With only days to spare, we had no alternative but to go ahead and use them.

The incoming models were assembled at Bawdsey and tested against our standard target, the Trimley water tower. Due mostly to the improvement in gain coming from the Pye receiver, the range performance was excellent. The complete assemblies were then taken to Martlesham for installation in the Blenheims, which were now parked in a long line on the aerodrome. This was no neat production line under cover, it was strictly installation among the daisies and the dandelions, completely exposed to the wind and the weather. It was a case of all hands to the pumps, and every one of us pitched into installing units, running them up on the ground and then flight-testing them. We were helped enormously in the installation task by five civilian fitters who were drafted in from Farnborough under the supervision of a friendly foreman, whose name I seem to remember was 'Charlie'. These proved to be pearls without price and they soon carried the main load of actually fitting our airborne radars to the aircraft. However, it was obvious that starting from nothing on 1 August, we would never complete 30 installations by the end of the month.

The completed aircraft were to go to 25 Squadron at Northolt, who knew nothing about operating the equipment they would shortly be presented with. So in addition to the fitting programme, we hastily set up a schoolroom in a wooden hut alongside 'D' flight at Martlesham. The radar sets would be operated by rear gunners of 25 Squadron and eight or ten of them, together with several pilots, came up from Northolt to be shown the equipment and to receive some elementary instruction on what radar and air-interception was all about—all in the course of a week or ten days. The first Blenheim was delivered to the squadron in mid-August and, thanks to prodigious efforts, five more were delivered by the end of the month. By any standards, this must be regarded as a minor miracle. Hanbury Brown went with them to give further instruction to pilots and crew and to help with maintenance. This was a monumental task which in normal times would have called for a whole team to accomplish, but we could only spare one man and Hanbury was the only one with the necessary experience. One of these Blenheims was the only radar-equipped night fighter to be in the air over London on the first night of the war, with a civilian—Hanbury Brown—in the back seat as radar operator.

As already explained, the maximum range of an air-interception radar set on metre waves was limited by the height of the aircraft above ground. At a flying height of 15000 feet this would be just under 3 miles. One of the first things which Hanbury discovered was that while the accuracy of the chain air-warning system was adequate for directing day fighters, it could not reliably put a night fighter within the above range of a target. There was an urgent need for a more accurate system, preferably one which gave an instantaneous indication of the relative position of the fighter and its target. I passed this on to the management of the Air Ministry Research Establishment (AMRE), a request which was backed by strong pressure from Fighter Command. There was an excellent response, and in a little over twelve months AMRE produced the so-called Ground Control Interception radar or GCI. This was a highly accurate radar which gave the operator an instantaneous range and bearing of both the night fighter and its target; with this information, an operator on the ground could tell a night fighter pilot exactly where to go to effect an interception. These GCI radars came into operation towards the end of 1940 and, as we shall describe, were a major factor in the final solution of the night interception problem.

We did not deliver the remainder of the Blenheims until later in September because yet another event—the declaration of war—totally disrupted our efforts and put paid to any hopes we had of bringing about a smooth and orderly introduction of airborne radar to Service use.

5

The Outbreak of War and the Move from Bawdsey

Several months before war was actually declared there had been an instruction from the Air Ministry that in the event of war Bawdsey Research Station should be evacuated to a safer site. An internal committee, with Rowe as chairman, was formed at Bawdsey to consider all aspects of such a move and to make recommendations. At that time, Rowe had a palatial office overlooking the sea and the senior staff had frequent meetings to discuss the problem. Many alternative sites were proposed—the most favoured being on the south coast or inland along the Severn valley. The choices were not always unbiased and I can well remember noting that there was almost a one-to-one correlation between the site suggested and the place where the person advocating it was born.

In an attempt to find a more rational choice, I drew maps based on distance of the suggested sites from the sea, availability of a good airbase, a good rail connection to London and so forth. These pointed to several favourable areas which were broadly along the Welsh border from Manchester down to the Bournemouth region. They may not have been the best places in Britain for a radar establishment but they were infinitely better than where we actually ended up. No one was in favour of going north of the Pennines. Rowe was curiously unwilling to make up his mind and no actual site inspections were ever made.

Most contentious was the projected date of the move—it was to take place on the first day of war. I pointed out that nothing could be more ridiculous. Roads and trains would be congested with evacuees streaming out of London, there would be urgent troop movements in every direction and there might even be bomb attacks. Although I was only three and a half years of age at the outbreak of World War I, I still had vivid memories of the congestion on the first few days of that war. The Bowen family had been on a short holiday to relatives at Fishguard and I can still remember the discomfort of a vastly overcrowded train journey on the Great Western Railway from Fishguard back to our home in Swansea. Thus, from a very early age, it was impressed on me that the first few days of a war were no time to be travelling if one could possibly avoid it. I argued that the right time to move was three months before the best estimate of a declaration of war or three months after. Preferably three months after, as it might then be found that there was no need to move after all.

In Rowe's office, with a glorious light coming in over a limpid sea, there was plenty of argument about when to move, but Rowe would have none of it. He was firmly convinced that Bawdsey would be a prime target, that it would be bombed out of existence on the first day of war and it would be folly to stay around. To me this did not gibe with the fact that we were the principal Air Defence Establishment in the country. We were supposed to know when the bombers were coming and where they were heading and we had already devised a credible system of sending fighters against them. If we were the first to cut and run, what about the fate of the rest of the population?

A decision on where to go was not made until Watson Watt appeared at Bawdsey at the conclusion of one of his regular tours of the chain station sites. He had been to the north of Scotland and on the way back had called at his old University at Dundee and made himself known to the Vice-Chancellor. He must have been overcome by nostalgia and remembrances of things past because he mentioned that there was a problem—in the event of war being declared, some members of an important Defence Laboratory might need emergency accommodation. 'No problem at all,' said the Vice-Chancellor gaily 'we would be only too glad to help in a case like that'—without enquiring too closely into how many staff were involved and how much space was required.

Like many things which happened in Bawdsey about that time, this was a matter which was not up for discussion—it was an accomplished fact. So Dundee it was. I asked Watson Watt about the airborne group and aerodrome facilities, which were simply non-existent at Dundee. He indicated that there would be no problem at all; during the same trip he had called at the civil aerodrome at Perth and made arrangements with the Station Manager for laboratory accommodation and hangar space for all our aircraft. He must have told a very convincing story because I accepted it without question.

So, the die was cast—Dundee and Perth it was to be. Plans were made to charter an entire train to transport the heavy laboratory equipment out of Bawdsey when the day arrived. The airborne group did not present much of a problem. We were few in number, most of the scientific staff had their own cars and much of our lighter equipment could be transported to Perth in aircraft of 'D' flight. The fateful day arrived and we were warned the night before—late on 2 September—that the trek north would begin the next day. Touch looked after things at Bawdsey and I went to Martlesham at the crack of dawn to supervise the loading and departure of our aircraft. By this time 'D' flight consisted of about 10 experimental aircraft but, in addition, there were 20 or more Blenheims destined for 25 Squadron. Many of them were half fitted with air-interception equipment and this made a total of over 30 aircraft.

Non-operational flying up the east coast was banned so the flight route to Perth was across country to the Irish Sea, then north to Glasgow and back across the Scottish Lowlands to Perth. The pilots of 'D' flight flew their own aircraft with as many ground crew as they could carry. In many cases they made several trips back to Martlesham to collect the necessary stores. For the Blenheims we enlisted the services of Ferry Command who had flown them into Martlesham in the first place. They were loaded with as much electronic equipment as they could safely carry and off they went. If this description of our departure sounds comparatively well organised, there were in fact some problems. Quite early in the morning and completely without warning, a stream of 10-ton trucks turned up at Martlesham loaded with soot of the blackest and stickiest variety. They started to spread it in large gobs over the aerodrome. Apparently this was something called aerodrome camouflage. If the enemy had decided to attack Martlesham Heath that morning, far from being obscured, its position would have been obvious from many miles away from the huge cloud of soot which went up in the air. Whoever thought that up for the first day of war must have been crazy. What with trucks running all over the landing area and aircraft taking off at frequent intervals, there was absolute chaos. Everyone was quickly covered in soot and we all looked like coal miners coming out of the pits. Luckily, the country's supply of soot must have been exhausted on the first day, because the exercise was never repeated.

Another event of that morning was that we had been told to listen to the radio at 11 AM when Chamberlain would broadcast the solemn facts about a declaration of war. Rowe later wrote 'At eleven o'clock on 3 September 1939, in a little cottage in Northumberland, I heard Neville Chamberlain announce that we were at war.' I listened to it standing outside one of the huts of 'D' flight with aircraft engines running up all around, already black in the face from the clouds of soot. Rowe was not noted as a fast driver and no one has ever explained how he got so far ahead of the field in so short a time. Meanwhile, back at Bawdsey, which he had repeatedly said would be wiped out on the first day of war, many of the staff were still sweating getting the equipment away.

It was dusk before I saw the last of our aircraft leave Martlesham and was free to attend to my own affairs. I returned to Felixstowe, picked up my wife who had packed a few bags, locked up the rented home—which I was not to see again—and headed north. Full black-out restrictions were in force and it was soon obvious that we would not be able to travel far. Near midnight, we put up at a guest-house just outside Ipswich. The proprietor is no doubt dining out on the story to this day. How, on the first night of the war, a character turned up demanding accommodation. He had a car, luggage and a perfectly clean and respectable wife, while he himself was filthy dirty and apparently covered in soot. British boarding houses are not well known for their bathroom facilities but he reluctantly agreed to my taking a bath, provided I did not use too much water. He will then relate how it took him a week to get the filthy marks off the bath, and how, apart from sending all the bed linen to the laundry, it took another week to get the soot out of the room we occupied. It was certainly several days before I managed to eliminate all that soot from my clothing. We left next morning, hell-bent for Perth, and I arrived a full 24 hours behind the advance guard. I was greeted by a peremptory message from the Superintendent at Dundee: 'Where the hell have you been? Don't you know there is a war on?' According to Rowe, I should have been right up there in the front line, running away from an enemy which had not yet fired a single shot.

After the flight of the research staff, Bawdsey remained one of the most important operational stations of the east coast chain. The Germans made occasional daylight attacks on radar stations on the south coast but, throughout the war, there was never a concerted attack on Bawdsey Manor. Three stray bombs dropped within the grounds at different times but there was no damage and no casualties. Such are the fortunes of war that the inoffensive and entirely undefended village of Orford, where we spent so many happy days in 1935, suffered a bomb attack in 1942 during which 13 villagers were killed and many cottages destroyed.

Of around 30 aircraft which left Martlesham on 3 September, one failed to appear at Perth. This was one of the Blenheims intended for 25 Squadron, flown by a pilot of Ferry Command. It was partly fitted with air-interception gear and carried a full load of additional components. Its fate was never properly established. It was presumed to have taken the same route as the other aircraft over the Irish Sea and that the pilot had run into trouble and gone down in the sea. There may be a more sinister explanation but, as far as I am aware, no questions were asked and no inquiry was held.

Dundee and Perth

When he arrived at Dundee, Rowe was greeted by an incredulous Vice-Chancellor who seemed to have forgotten Watson Watt's visit; he had simply not been informed of any subsequent plans. Being an accommodating soul, he offered the new arrivals two rooms, each 20 feet square, which was all the space he could spare for the several hundred people who were either *en route* to Dundee or already there. I do not want to elaborate on how the problem was finally solved and will leave those responsible to explain what transpired as best they can. There had been a monumental blunder, from which the Air Ministry Research Establishment, as it was now called, took a year or more to recover.

Our own reception at Perth was not much better. The Station Manager was a sensible and friendly character but he made his position very clear. He was a civilian under contract to Training Command of the RAF to provide *ab initio* flight training to young pilots. This was the usual way of conducting early flight training in the RAF and there were many such centres around the country. His contract had been in force for twelve months or more and, as far as he was concerned, was likely to continue for the remainder of the war. His programme had grown to the point where it required the full facilities of the station and there was nothing to spare. He remembered Watson Watt's visit, but there had been positively no follow-up. Neither he nor his masters had been told what was going on.

After a few telephone calls to the Air Ministry by both of us, the Station Manager offered a friendly compromise. Until the matter could be resolved, he would share equally the office and hangar space which was available. There were two small hangars and associated office and work space and this was an arrangement to which I readily agreed. He took one hangar and I took the other, and for the moment I was grateful. It meant that we had a tiny space for laboratory work and equipment, and hangar space for one or two of the aircraft actually being worked on. The remainder were parked in the open, partially fitted with delicate electronic equipment exposed to the weather. We talked to our respective masters at the Air Ministry and the message came through, loud and clear. Perth was considered an ideal place for *ab initio* pilot training and totally unsuitable for airborne radar research. Who had been silly enough to think that the airborne group should go to Perth? It became obvious that the Station Manager would shortly get his station back and that another place would be found for us.

Conditions at Perth were rudimentary compared with those we had left behind at Bawdsey and Martlesham. The few rooms available to us were never intended as laboratory space and in many cases were simply areas which had been partitioned off for some other purpose in the corner of the hangar. Things were terribly congested, but we were told there was a war on and that this was the kind of thing one put up with. The personnel of 'D' flight were much worse off than the scientific staff. The long-suffering ground crew did all their work in the open and in the case of bad weather they had no alternative but to congregate in the hangar. Flight crew were no better off; there was a small room in the main airport building which, in peace time, had been the public lounge and reception area. Between flights, this was where the aircrew congregated with their flight gear, their parachutes and the rest of their personal belongings spread around the floor.

Perth was still rated a civil airfield, so there was virtually no security. In my office, we regularly marked everything 'Secret' or 'Top Secret', as the case may be. We religiously kept the filing cabinets locked when not in use and locked the doors when we left, usually very late at night. There were plenty of Air Force personnel around in daylight but only a perfunctory watch was kept at night. There was no guard of any kind around the airport perimeter. There was top secret radar gear in the aircraft scattered all over the field and in the rooms used as laboratories. Almost anyone with the initiative could have walked in and got away with a complete set of the country's most secret radar equipment.

Not that this was anything unusual. Martlesham Heath had been the nation's top aircraft testing establishment. In the five years before the war, all the newest aircraft going into service were tested there, putting in intensive flying, day after day. Not one but two main roads went through the camp—through it, not around it. It is true that after landing the most secret aircraft were hurried under cover as soon as humanly possible, but many of them had to taxi across a main road before doing so. This was true of every one of our radar-equipped aircraft. Bristling with unusual aerials, each one of them had to taxi across a main road before taking off and do so again after landing. It had always been like that and was accepted as normal. The day after the declaration of war, when all the secret aircraft had gone, a contingent of WAAFs took up residence at Martlesham. Both roads were promptly closed. Whether this was to keep the WAAFs in or to keep intruders out is a moot point. Whatever the reason, those WAAFs must have been flattered to know that all of a sudden the country was much more concerned about protecting their virtue than it had previously been about safeguarding Britain's most precious secrets.

Meanwhile, at Perth there was an immense amount of work to be done. Our first priority was the installation of AI radar in the remaining Blenheims and, thanks to superhuman efforts from the staff, this was completed in October. At the same time, Coastal Command had come into the picture and asked for the installation of ASV radar in Lockheed Hudsons, to be followed by flying boats. In addition, the Fleet Air Arm asked for an experimental fitting of forward-looking ASV radar in a Swordfish and a Walrus, both of which shortly arrived in Perth.

Anything remotely resembling research or development had to take a back seat. All the emphasis was now on getting sets produced and installed in Squadron aircraft. Among the first things to be fixed was the final design of ASV Mark I. A couple of days after the outbreak of war, I was back in London for urgent discussions with the Air Ministry about the next steps in that programme. The E K Cole Company was brought in as contractor for the next batch of airborne transmitters. These, at last, were to be up-to-date models in place of the antiquated versions we had suffered from up to that time. We settled on the design of an AI Mark II which still used DET 12 and TY 150 valves in the transmitter, and the ASV Mark I design, which became the first airborne set to incorporate the new micropup VT90 valve produced by GEC at Wembley.

An unusual event took place during that visit to London which had important consequences. I was approached by two members of the ICI Company who carried samples of a plastic material which they had just produced at their factory in Warrington. It was a flexible material, rather like rubber, which they thought might have good radio-frequency properties. They asked if it was likely to be of any use as a filling material for concentric cable. I seized on the idea and took samples back to Perth for radio-frequency tests. This, if it came off, would solve one of our most difficult problems. Up to that time all our radio-frequency cable runs were literally made by hand from small plastic beads threaded on a flexible inner conductor and then enclosed in a braided exterior. It was a slow and laborious process and the resultant product was frail and easily damaged while being pulled through the confined spaces of an aircraft.

The new material proved to have excellent radio-frequency properties. ICI soon went into limited production and made sufficient cable to satisfy our immediate needs. It was vastly superior to anything previously available and immensely simplified the problem of installing radar aerials on aircraft. The material was of course polythene, which is now in world-wide use for an enormously wide range of applications. To us it was a magical material which came like a heaven-sent blessing. Airborne radar was the first to benefit from it. The raw material was not easy to produce but, thanks to a gigantic effort by ICI, it soon spread throughout the radio and radar industry.

The request for ASV installations for the Fleet Air Arm had a most interesting outcome. When the Walrus finally arrived at Gosport, the radar was run up on the slips and demonstrated to the Admirals. It was only about 20 feet above sea-level, but from this limited height it gave echoes from a whole array of ships spread over the Solent. Mountbatten was in attendance and he was so impressed he ordered one of these radars to be fitted to his destroyer *Kelly*. Using ASV components, it became the Naval Type 286 and 200 of them were subsequently ordered for fitting to destroyers and torpedo boats.

Back at Perth, we were now getting a small number of additions to the staff. Early in the war, plans had been made to recruit scientific people from the universities directly into radar. They later played an outstanding part in the whole development of wartime radar. It took them some time to get used to the rituals of government service, and this was not made any easier by the survival conditions in which we found ourselves. Moreover, they were not always allocated the most glamorous of jobs. For all that, they took to the task with a will; what is more, they were fresh and not suffering from the physical and mental strain which the pressures of the previous few months had brought to many of us. The first such recruit we received into the airborne group was A C B Lovell from the University of Manchester, later one of Britain's most distinguished radio astronomers, and we were blessed with several more in the months which followed.

It was a time when our resources were strained to the limit and we were under tremendous pressure to get AI and ASV radar into fighting aircraft, but a minor incident took place at Perth which illustrates the loads which were sometimes thoughtlessly placed upon us. Watson Watt was not an infrequent visitor and called in several times during his many site inspection trips. He sympathised with the difficult conditions under which we had to work and was clearly intent on finding us another base. He was just leaving after one of these visits, I had in fact walked with him to the exit where his car was parked, and was just about to step in to the car when he said: 'Oh, by the way, the Navy want some instruction on airborne radar. They are sending a group up on Monday for a two week course. You will look after them, won't you?' Then he was off in a cloud of dust. I believe it was a Thursday and I had the weekend to think about it.

In the event, 20 ensigns and ratings showed up on the Monday morning with a Naval Commander in charge. We had practically nothing to offer—not even a room to accommodate them. Luckily, he was a very experienced professional and was prepared to improvise. He gave me the impression that he had met this kind of situation before. We quickly agreed to a simple programme. During the first part of the morning his men would be free to move about the laboratory space looking at any equipment which was in use and asking questions. Mid-morning there would be a single lecture lasting an hour, followed by another hour of question time. I gave the introductory lecture on the general principles of radar, and this was to be followed in turn by staff members who covered pulse transmitters, receivers, indicators and so on. After the lecture they were free to inspect the aircraft and examine those with equipment being run up. We were far too heavily loaded to offer a proper programme of demonstration flights but they were free to participate in any test flights being made. I left it up to the Commander to watch the flight roster and arrange what he could.

The afternoons were free for the Commander to arrange as he wished, and he certainly did not let the grass grow under his feet. He had his men square-bashing for several hours, followed by soccer and rugby games on an adjacent field. If those sailors did not learn very much about radar, they certainly arrived back at their units in fine physical shape. In the end it probably worked out reasonably well, but I put it down entirely to the goodwill and good sense of that Naval Commander. This was no way to perform the teaching and training of radar personnel; it was a long time before it was realised that the only way to proceed was to organise proper teaching courses staffed by well qualified professionals.

During the two months we were at Perth, I visited Dundee every couple of weeks, mostly to arrange administrative matters like payment of salaries and the purchase of supplies. If we had problems at Perth, it was nothing to the malaise which beset the main establishment. My recollection of Dundee is of completely dispirited people, sitting around wondering how to occupy their time. They were crowded into inadequate accommodation, with very little equipment with which to work. Most of the work which mattered was going on at field stations, not at Dundee itself. Compared with Bawdsey it was a very depressing picture. I could find no one interested in discussing technical matters. Rowe was overwhelmed by domestic problems and had no time to discuss any difficulties we might have at Perth. It was the beginning of a period of almost total neglect of the airborne group by the main establishment; it went on for a long time and did nothing for our morale.

The Move to St Athan

Towards the middle of October, we were told that our new home would be at 32 Maintenance Unit, St Athan. This was in South Wales and, as it happens, about 30 miles from where I was born. This time the rationale for the choice of St Athan could not be faulted. The Unit had been established principally for airframe and engine maintenance but it was now given the additional task of specialising in the installation of all classes of airborne radar. Our principal job was to instruct the Unit in the fitting and testing of airborne radars, after which we could return to our proper sphere of research and development. It was never intended that we should stay at St Athan longer than was necessary to perform this task and the expectation was that when it was completed we would move on elsewhere. This was not always appreciated by our newly arrived appointees from the universities who did not always understand that, for the time being at least, we were unable to carry out any research and development.

On paper, St Athan had much to recommend it. It was one of the many huge establishments built by the Air Ministry in anticipation of the war. It had a large landing area with one exceptionally long runway. On the north side were extensive school and training buildings. To the south was 32 Maintenance Unit, with several C-type hangars and large machine shops which were unusually well equipped. Alongside was mess and barrack accommodation, all well built and beautifully proportioned. The trouble from our point of view was that with new recruits pouring into the Air Force in great numbers, there were 4000 people on the station and it was already full to overflowing.

I was met on arrival by the Commanding Officer, Group Captain Lucking, and his Chief Engineer, Wing Commander Fowler. They had already been briefed by the Air Ministry on their part in the task, but judging by their reactions I think they expected us to be a small group of itinerant fitters and riggers, with a few boxes carrying our tools of trade. In terms of the task in hand, namely responsibility for every aspect of airborne radar, we were still a small group made up of scientists, technical assistants and other ranks; a total of 25 people. We were far outnumbered by 'D' flight, which consisted of about 40 officers and men. The flight itself comprised a permanent establishment of 8 or 10 experimental aircraft but, in addition, there was a flow of aircraft coming in from Squadrons day after day. When we arrived at St Athan I got the distinct impression that the Commanding Officer had not been informed of the true magnitude of the operation.

At considerable inconvenience to themselves, the Unit agreed to make one C-type hangar available to us, but they made it clear this was the absolute limit. St Athan was a permanent station and there was an Air Ministry prohibition on building temporary structures of any kind within the camp. Any building project had to go through the full

Department of Works treatment and, even if it were approved, such plans would take at least twelve months to implement—by which time we would long since have gone.

Superficially, a C-type hangar provides a lot of space, but it is space suitable only for filling with the larger variety of aircraft. With the doors permanently open, or at best opening and shutting frequently, the interior can be the coldest and windiest place on earth. The hangars had nothing remotely resembling laboratory space—only a series of open annexes around the walls, wide open to the wintry blasts blowing through. The plan was to partition these annexes off to make laboratory space, but we had no funds for this and neither did St Athan. Wing Commander Fowler did what he could to help. He produced vast sheets of ready-doped aircraft fabric and suggested these be placed over a framework of battens to act as partitions. The material was a camouflage green colour which looked strangely familiar. It turned out to be the outer skin of Handley Page Heyfords—the type of aircraft we had used at Martlesham Heath for our very first airborne trials. The Heyford had been obsolete for years but it transpired that somewhere in England a factory was still turning them out at a rate of several a week. There was no possibility whatever of their being used in the present war but someone had forgotten to turn off the tap. The Heyfords were flight-tested at the factory and then flown straight to St Athan, which had the unenviable task of stripping them down to spare parts. The Rolls Royce Kestrel engines were about the only things worth salvaging and these eventually found their way into a training aircraft which was specially designed to use up these beautifully built, but outmoded, engines.

We might have got by with those canvas partitions if it were summer, but it was already November and for several months there was no source of heat whatsoever. For a large part of the time our people wore greatcoats and gloves inside this improvised laboratory space. The sight of a future University Professor, in coat, gloves and cap, working on a radar chassis with a soldering iron will stay with me for the rest of my life. The conditions were simply appalling—and any member of the airborne group will say that this was by far the most unpleasant and least productive part of their lives. Here was one of the most sophisticated of defence developments being introduced to the Royal Air Force, and it was being done under conditions which would have produced a riot in a prison farm. I often wonder how much better the night defence of Great Britain would have been in the latter half of 1940 if we had been provided with better facilities and greater manpower during those critical days at Perth and St Athan. It is easy to be wise after the event, but I blame it all on that precipitate flight from Bawdsey; responsibility for this must be placed firmly on the shoulders of Rowe and Watson Watt. They had the task of deciding what to do and where to go—and they so decided. As was already abundantly clear, Bawdsey and Martlesham had not been singled out for special treatment by the enemy and, once this was obvious, I will never understand why we were not allowed to go back.

If only to underline the irony of the situation, one miserable morning, a few weeks after our arrival at St Athan, conditions were not the best for flying and very few aircraft were in the air. An aircraft appeared out of the east and seemed at first to be making a normal approach. However, several things were wrong. It was going much too fast and the wheels were up. It headed down the main runway and levelled off at a few hundred feet. We could then see that it was a Ju88 with its bomb doors open. A 1000 pound bomb came out of the bomb bay and those who saw it ducked for cover. There was an agonising pause—but no explosion. In fact, the bomb bounced several times down the runway like a tennis ball and buried itself in an embankment at the far end of the field. It was subsequently recovered by a bomb disposal squad, who found that it was a delay bomb—the pilot would not have dared drop it otherwise—and the detonator had failed to work.

The Task at St Athan

In spite of the conditions, work at St Athan fell into a pattern. Pye Radio and E K Cole were now firmly established as the two principal manufacturers of airborne radar. We had frequent design talks with them in London. These could hardly be called design conferences—they were often held on uncomfortable chairs outside someone's office at the Air Ministry or late at night in hotel lounges after a full day's work elsewhere. In a remarkably short space of time, deliveries started coming in. Our first task, as at Martlesham, was to assemble complete units on the bench and test them against the usual ground targets. Those which tested out well, and most of them did, were passed to the installation crews who made fittings and cable runs and then installed them in the designated aircraft. Next, the sets were run up on the ground and, when up to standard, would be flight-tested. The test flying was done by pilots of 'D' flight, with Mills, Chalky White or Keith Wood doing the final flight checks. The aircraft were then flown out to the appropriate Fighter Command and Coastal Command stations.

For a start, we did most of these tasks ourselves but, gradually, they were handed over to St Athan personnel. The process usually started with a Flight Sergeant—a canny, well trained and very experienced fitter/mechanic type who quickly picked up the fitting routine. He would assemble a crew to take over the fitting task and gradually take on full responsibility. As this team became skilled, they in turn passed their skills on to new trainees, and from quite small beginnings a strong work-force was built up. One could not fail to be impressed by the way those early Flight Sergeants took hold of the situation. They were mostly men with long terms of service, wise in the ways of the RAF and masters of getting things done with an inexperienced work-force.

The process was slow at first and did not get properly under way until the middle of our first month at St Athan—November 1939. The momentum built up and in December, the first full month, the fitting record was:

AI in Blenheims—17
ASV in Hudsons—3

As the fitting crews gained experience, the January figures were better:

AI in Blenheims—18
ASV in Hudsons—12

That is, during the first month of 1940, aircraft fitted with either an AI or an ASV radar were being delivered to squadrons at a rate of one a day. The figures improved during succeeding months and the total number of such aircraft delivered from St Athan in the eight months up to 31 July 1940 was as follows:

AI in Blenheims—118
ASV, mostly in Hudsons—91

These were the figures for St Athan alone, but this was not all. Where aircraft could not be delivered to St Athan, for example Sunderlands and Catalinas from Pembroke Dock and Helensburgh, fitting parties went out to work in the field and many additional aircraft were completed in this way.

A most important task, begun at St Athan and continued elsewhere for another year or two, was the design and installation of sideways-looking radar on Coastal Command aircraft. These were operational versions of the aerials first used for sideways-looking radar, as described in Chapter 3. They consisted of broadside transmitting arrays on top of the fuselage of large aircraft like the Sunderland and the Wellington, and further arrays along the fuselage for receiving from the port and starboard respectively. Examples of these are shown in figures 6.2, 6.3 and 12.3 and they were destined to play a significant part in the battle against submarines. The work was arduous and was different for each aircraft, involving a long period of measurement in the field, followed by verification in the air. Much of this work was carried out by Hanbury Brown and Keith Wood and it kept them occupied long after our departure from St Athan.

The airborne group's time at St Athan was drawing to a close, but the fitting programme continued as the main responsibility of 32 Maintenance Unit. St Athan personnel trained three other Maintenance Units to perform similar tasks and, before the end of the war, the rate of delivery of radar-equipped aircraft to operational squadrons approached one hundred a month. The record of this achievement is buried deep in the Air Ministry files; what those Units did is a vital part of the radar story but an adequate account of it has never been written until now.

Although we were still a section of AMRE at Dundee, the airborne group was virtually without back-up from that source. Our technical discussions were with people like Watson Watt and his colleagues at the Air Ministry and with the manufacturers, E K Cole and Pye Radio. Design and production arrangements were made with them directly, without reference to Dundee, while the complex arrangements for moving aircraft to and from the squadrons and fitting their radars were made during frequent visits which I made to Fighter Command and Coastal Command. I asked repeatedly for an administrative officer, or even a clerk, to supervise some of the more routine matters and we knew of several old friends on the administrative staff at Dundee who were dying to get away—but the management at Dundee could see no merit in this arrangement. People like Watson Watt, Tizard and Blackett found time to visit us at St Athan, mostly to review technical progress, and all of them were appalled by the conditions under which we had to work. Early in 1940, Tizard paid a special visit to Dundee to intercede with Rowe about his neglect of the airborne group, but this brought no measurable improvement.

The Royal Visit

However, we were not entirely without official recognition of the task we had in hand. On 9 February 1940 the station was honoured by a visit from King George VI and Queen Elizabeth. In the previous few days, intensive preparations were made for the visit. We cleaned out the hangar and lined up an imposing array of aircraft being fitted with the different types of radar. The Air Force was busily engaged in cleaning and polishing the aircraft when, about mid-afternoon, only 24 hours before the Royal visit, there was an ear-splitting 'Rat-at-at-at-at' through the hangar and .303 bullets sprayed everywhere. Luckily, no one was hurt, but there were some dreadful scars around the walls and on several of the aircraft which had been lined up.

When order was finally restored, we discovered what had happened. Among the aircraft on display was a Wellington which had just returned from operations. An aircraftsman, sitting in the front gun turret—a four-gun turret—was amusing himself swinging it around. He imagined he was in action against the enemy with a target in his sights; he pulled the trigger and found to his horror that the guns were loaded. They were pointing dangerously near the path to be taken by the Royal party. Needless to say, it was not long before all the aircraft in the area were stripped of their ammunition.

It was far into the night before I finally retired to my hotel room at Southerndown a few miles away. I may have got three or four hours sleep that night, after which it was back on the road again for the Royal visit. At that time I owned a Riley 9—these days a delightful vintage car—of which I was very proud. Driving back down the road to St Athan that morning, overcome by weeks of overwork, I fell fast asleep at the wheel. Apparently, I drove off the road and the car slid over on one side in a sea of mud. Miraculously, before finally coming to rest, the car came right-side up again. I sat in my seat for a few minutes, confused but unhurt, and then pressed on for St Athan. I was relatively unscathed but my treasured black and white Riley bore the marks of the encounter until the end of the war.

The Royal visit was an impressive function and went off without a hitch. The Air Force had their own line-up on the tarmac, complete with brass band, while the airborne group stood in line inside the hangar for the Royal inspection. We had been warned that the King might show signs of a stammer and that we were to pay no attention to it. In the event, he gave no sign of this whatsoever and, to my surprise, showed himself to be quite well acquainted with radar, both in its ground and airborne applications. Apparently, he had discussed the whole subject with Watson Watt during a surprise visit to one of the chain stations in Yorkshire a few months earlier. He appeared well pleased with what was going on. He passed down the line and had a pleasant word for every member of the group. We were honoured by the visit and it gave an enormous boost to a group which had been feeling the cold, cold winds of adversity, in more senses than one.

6

ASV Radar Against Ships and Submarines

As we have seen, Coastal Command had been rather slower than Fighter Command in asking for airborne radar to be fitted to their aircraft, but a start was made at Perth when we were asked to finalise the design of a Mark I ASV and prepare it for installation in Lockheed Hudsons, and in a Swordfish and Walrus of the Fleet Air Arm. By the time we arrived at St Athan, production models of ASV Mark I began to arrive from the manufacturers and were put through the regular tests prior to installation. The first squadrons to be fitted were 220, 224 and 233 Squadrons. Interestingly enough, 220 Squadron was where our first two Ansons came from back in 1937. Now equipped with Hudsons from the USA, the Squadron had moved to Leuchars in Scotland and was heavily involved in sea-search and escort duties associated with the Norwegian campaign.

The ASV radar was an altogether easier equipment to install than the AI. The Hudson was a roomy aircraft so there was no space problem. The maximum range of detection on a capital ship was now more than 20 miles, so that a slight drop in performance was not likely to be serious. The Hudsons were flown in from the squadrons and by the early part of 1940 we were delivering two or three a week. Things were getting into a routine and there were far fewer problems than with the first Blenheim installations back in August.

We had two valuable new recruits in John Pringle and Alan Hodgkin. Hodgkin was an enormously talented physiologist from Cambridge who, after the war, earned the Nobel Prize, became Master of Trinity College and was a President of the Royal Society. He became a driving force in the development of centimetre-wave air-interception radars in Britain. Pringle was also from Cambridge and had already made a name for himself as a zoologist. He was especially welcome in the airborne group because he was a skilled glider pilot who already had that most important quality—a feel for what could and could not be done in the air.

About this time things had progressed well enough for us to send a small team to Pembroke Dock, a flying boat base from which 10 Squadron of the Royal Australian Air Force was operating Sunderlands over the Bay of Biscay and the Atlantic approaches. Pringle took charge of the work on the Sunderlands and it was not long before the whole Squadron was fitted. Work then started on the Catalinas, the long-range flying boats from the Consolidated Aircraft Company in California. As each new squadron was fitted, the aircrews had to be taught the rudiments of operating the equipment and how to make best use of the search information provided. Hanbury Brown and Keith Wood did this for land-based aircraft and Pringle for the flying boats. Our principal contact at Coastal Command HQ was Squadron Leader Lugg who was a tireless worker and kept things moving at a good pace.

It was this officer who had the brainwave of modifying an IFF set and installing it in the control tower at Leuchars as a homing beacon for the ASV. This calls for some explanation. IFF means the identification of friend from foe and was an early product of the work at Bawdsey. It was a small device, which when carried in one of our own aircraft sent out a signal which allowed a chain station to identify it as friendly. It was simply a receiver which picked up the signal from a chain station and returned it in amplified form; a coded signal was added which identified the aircraft carrying it as friendly.

The scheme which Lugg suggested was exactly the reverse of this; that is, the IFF set in slightly modified form was usually installed in the control tower of an airfield. A radar-equipped aircraft could then interrogate it and obtain both a range measurement and a bearing on its home base. This was a marvellous innovation, and was particularly helpful at Leuchars, where there was high ground just west of the landing field. Returning from patrol over the North Sea in bad weather, aircraft would sometimes overshoot their destination and there were several cases of crews being lost by collision with high ground. The beacon was a resounding success and an ASV-equipped aircraft could obtain an accurate range and bearing from its base when it was still 50 or 60 miles from home. For obvious reasons, the beacon became known as 'Mother' and it saved a significant number of lives during the remainder of the war. Its use quickly spread to other squadrons and to Fighter Command, where it became an integral part of the interception procedures. Thinking back over the number of times we had felt our way back to Martlesham in bad weather before the war, I cannot understand why we did not think of such a beacon then. However, the honour belongs to Squadron Leader Lugg—and I doubt whether he has ever received full credit for something which was an enormous help to operations and undoubtedly saved many valuable lives.

By far the most important task of Coastal Command in those days was in providing an escort for convoys; ASV radar proved an invaluable tool for finding the convoys at night, or in bad weather, from a range of 30 miles or more. This, together with the comfort provided by the radar beacon which allowed them to home in on their base, was probably the feature which first endeared Coastal Command crews to the equipment.

There are not many records from that period of Coastal Command using radar to attack enemy Naval vessels. Most of these ships were hugging the coast of Norway or the coast of France and it was difficult to discriminate them from the coastline itself. But predictably, greater success could be expected from carrier-based Swordfish of the Fleet Air Arm operating in more open waters. In their hands, ASV radar played a conspicuous part in finding and shadowing the Italian Fleet just before and during the Battle of Matapan in March 1941. About noon on 27 March, an RAF flying boat first found that three Italian cruisers and several destroyers had put to sea in the Mediterranean; it placed them about 300 miles west of Crete and moving in a south-easterly direction. The Mediterranean Fleet closed in from Alexandria and contact was made next morning when Swordfish from the aircraft carrier *Formidable* spotted them and gave a new position. Fresh reports from search aircraft continued to come in during the day, including an additional spotting of the battleship *Littoria Venetu* about sixteen miles north of the cruisers.

About 3 PM, aircraft from the *Formidable* carried out several torpedo attacks and these succeeded in slowing down the *Venetu*. About sunset on 28 March, similar attacks were made on the cruisers, which were now identified as the *Zara, Fiume* and the *Palo*. The *Palo* was stopped in her tracks and, after darkness fell, the *Zara* and the *Fiume* returned to provide assistance. The main Fleet caught up with this group about 10.30 PM that night, after which a radar-directed air and sea attack was mounted. As a result, all three cruisers and several destroyers were sunk by gunfire and torpedoes. The *Littoria Venetu* escaped in the confusion, but in this single action three heavy cruisers and several destroyers were lost with upwards of 2000 men. This was the first big radar Naval battle of the war and it practically wiped out the Italian Navy as a maritime force.

It was only after this action that the German High Command realised that British ships and aircraft carried radar of considerable potential and that this made them a powerful night attacking force. It is a matter of historical record that the German High Command informed the Japanese of this fact and provided them with technical details of German radar, together with as much detail as they could assemble about British radars.

A few months later, ASV-equipped Swordfish played a crucial part in the search for the *Bismarck*. After she had sunk the *Hood* in mid-Atlantic on the morning of 25 May 1941, the *Bismarck* was engaged by the *Prince of Wales* and damaged by gunfire. She managed to escape in mid-Atlantic and for several vital hours contact was lost. She was found at 10.30 AM on 26 May by a Catalina of 209 Squadron of the RAF; she was then 700 miles west of Brest and seen to be heading for safety at that port. The cruiser *Sheffield* was called up and took on the role of shadower while ships of the Fleet converged from all directions. The weather was exceedingly bad, but late on the same day the air search was resumed from the aircraft carriers *Victorious* and *Ark Royal*. A single Swordfish could not carry both a radar and a torpedo, so the tactic was for the search aircraft to be accompanied by one or more attack aircraft carrying torpedoes. Towards nightfall, a search team from the *Ark Royal* found a likely target near the expected position of the *Bismarck*. The attack aircraft was directed on course, told to duck out of cloud and let go a torpedo. Sure enough, as indicated on the radar, there was a target directly ahead and the torpedo was fired. Seconds later, the pilot realised his mistake and held his breath in horror—his target was the cruiser *Sheffield*, hard on the heels of the *Bismarck*. The torpedo sped on and hit the *Sheffield* a resounding thump amidships. By an astonishing piece of good fortune, it was a dud and failed to explode.

The attacking aircraft did not get away unscathed and was damaged by anti-aircraft fire. However, it limped back to the *Ark Royal* and landed safely. Meanwhile, the search aircraft continued on its way and called up another torpedo-carrying aircraft. Miraculously, they were successful in finding one another and the search continued. There was low cloud and the light had gone when at about 9 PM they came upon another target. It had to be the *Bismarck*. Another attack was made and this time the torpedo struck her near the stern and put the steering gear out of operation. This was the beginning of the end of the great ship. She went around in circles and the following morning she was crippled by gunfire from the *George V* and the *Rodney* and finally sunk by torpedo.

First Trials of Air to Surface Vessel Radar against a Submarine

It is a remarkable fact that, until war was actually declared, little, if any, thought had been given to the use of ASV radar against submarines.

This situation changed dramatically towards the end of November 1939 when I received a long telephone call from Admiral Somerville at the Admiralty in London. It was most unusual for the Admiralty to call an Air Ministry establishment in this way, but the matter was clearly urgent and it was the way things were done in those hectic days. German submarines had, of course, been active since the very beginning of the war and shipping losses were beginning to mount. Around the coast of Britain regular Naval patrols kept them submerged by day but at night the submarines could not be seen on the surface and they roamed at will. It disturbed the Admiralty to find that the submarines were getting exceedingly bold and cruised about freely on the surface at night, in the English Channel and in the waters around Scapa Flow. This posed a serious threat to vessels of the Fleet itself. Could we do anything with ASV? Admiral Somerville made it clear that the submarines would make a very poor target. They would be running hull-down, with only the conning tower and very little of the hull showing above the sea surface.

This was the first time that submarines had seriously been considered as a target for airborne radar. Up to that time we had always thought in terms of larger vessels, the larger the better. I could not answer his query offhand but thought we might be able to detect submarines at a range comparable with that of aircraft, that is, at two or three miles. Admiral Somerville thought that this might be just good enough—could we arrange to do some trials?

He called back to say that a submarine could be made available for tests in the English Channel in a few days' time. I agreed to take a Hudson of 220 Squadron, which had just been fitted at St Athan, down to the Naval Air Station at Gosport for trials which would start on Saturday 2 December. Admiral Somerville aranged a meeting at Naval HQ in Portsmouth for 8.30 AM on that day to work out the details. Our plan was to fly down to Gosport the night before, with Hetty Hyde as pilot and Hanbury Brown and myself to handle the radar. As usual, the weather was atrocious and Hetty had some doubt whether we would get through, so at the last moment I decided to go to Portsmouth by car to make certain of being able to attend that 8.30 AM meeting. I started off as dusk fell, heading first for the Severn Ferry and then through the winding roads to Portsmouth. The route is difficult enough in broad daylight but at night, under black-out conditions, it proved nearly impossible. As I passed through Radstock, I found that the cloud was right down on the Mendips and it was clear that the Hudson would not get through. In the event, Hetty Hyde and Hanbury found that the cloud was right on the deck over Somerset and, with no certainty about what conditions were like at Gosport, they very sensibly turned back to St Athan. Meanwhile, I kept going through the night. Luckily, the roads were still signposted, but I frequently found myself, at a crossroad, having to stop the car and shin up the signpost in order to read it. More by luck than judgement, I muddled through and well after midnight was lucky enought to find a hostelry open and doing business at Fareham, just outside Portsmouth, where I found a bed for the remainder of the night.

The Hudson did not arrive at Gosport until later the next morning so I went alone to the meeting at Naval HQ. Admiral Somerville was there with members of his staff and the Captain of Submarine L27 which was to do the tests. Admiral Somerville outlined the seriousness of the problem confronting the Navy. As he had already explained, there was not much of a problem by day when Naval patrols could spot submarines on the surface. The submarines therefore remained submerged during the day, although they could still be tracked with underwater sound-ranging equipment—the so-called 'Asdic'. But at night they could not be seen by eye and the Asdic was ineffective against a vessel on the surface. This left the submarines free to travel at will during the night and this posed a serious threat to vessels operating out of the Solent and Scapa Flow. A submarine is most vulnerable when on the surface. If an aircraft could attack it in darkness, even if no immediate damage was done, this would be sufficient to make the submarine go down for a few hours. If, in this way, the submarines could be forced to submerge frequently by night as well as by day, this would seriously restrict their movements and significantly reduce the threat.

So plans were made to send Submarine L27 into the middle of the English Channel next day and for our aircraft to carry out a series of runs on it to determine the range of detection. The organisation necessary to carry this out proved unexpectedly complicated. As a protection against attack from enemy submarines, all vessels in open waters were required to change course every 20 minutes, either to port or to starboard in a random sequence. This made things difficult for us. We were accustomed to doing runs on a vessel ploughing a steady course, in which case it was relatively easy to go out of range and then come back again, either broadside or end-on to the target vessel. If the target were continually changing course this would be exceedingly difficult. We compromised, therefore, as follows. The submarine would only head in the four cardinal directions, north, south, east and west, in random sequence. At the conclusion of each run, as the Hudson came overhead, the submarine would signal its next course by Aldis Lamp. This would give us the clue we needed and we would head off for the next run. This would be repeated as many times as was necessary to give the needed range information. We then agreed on a

Verey pistol signal—two greens to start the exercise and two reds from the Hudson to end it, and a similar signal from the submarine if for any reason it wanted to terminate earlier.

Quite late that night Hetty Hyde, Hanbury and I paid a courtesy call on the L27 which was berthed at Fort Blockhouse at the entrance to Portsmouth Harbour. The Captain welcomed us in the wardroom—which was an exceedingly tight squeeze for four people—and we confirmed the signalling arrangements for the next day. We also fixed a map reference in mid-Channel for our rendezvous and a starting time, which I seem to remember was 9 AM. We were then shown around the submarine. It was my first experience of being below in a submarine and, although we were tied up firmly alongside the wharf, I must admit to awful feelings of claustrophobia. I made a firm resolve. Aeroplanes flying at 25000 feet and 300 miles an hour were perfectly alright by me, but 200 feet down in a submarine was simply not on.

The L27 had just come back from a 30-day patrol in the Heligoland Bight. The routine was a month on patrol followed by a month off if they were lucky. For that month they stayed on the bottom of the Bight by day, keeping completely quiet and listening out for any activity in the area. After dark, they would surface and observe what was going on. If there was enemy activity, they might be able to take a pot-shot at something, after which they would certainly be hunted for several days and might even have to abandon the patrol. If there were no activity, they would surface and recharge the batteries for a few hours. It must have been a ghastly existence.

We were shown the engine room, the battery room and the torpedo tubes, all marvels of compression, with practically no room for human beings anywhere. I was particularly struck by the radio room. This was a tiny cabin about six feet by four feet, stuffed with radio equipment of every kind. This was December 1939, at the start of what promised to be a war in which the best equipped side would win. The transmitter was a spark transmitter, inherited from World War I, with large copper coils and condenser plates like frying pans. It had been converted to valve operation, probably around 1921, and so it remained to fight a 1940s war. If anything, the receiver was even more frightening. It consisted of a row of DEQ valves stuck on a long sheet of ebonite. I do not have to describe it further; anyone far advanced in years will know what I am talking about. My first introduction to wireless, as it was then called, was literally as a child of eleven in 1922. The DEQ valve was already obsolete and, although I was familiar with it in photographs and diagrams, I had never actually seen one in real life. Here, I was seeing them for the first time on the main receiving panel of one of His Majesty's submarines, 20 years out of date! I had already, in the previous hour or so, generated an enormous respect for the British Submarine Service and what they put up with, but this was something else again.

I was subsequently told that when in the mid-1930s it became necessary to put out a contract for more of these receivers, the Marconi Osram Company pointed out that the DEQ had been out of production for 15 years and that valves of much better performance were available at a much lower price. They were told to have another think. A uniform product was of utmost importance to the Admiralty—would the Marconi Osram Company please get back into production of DEQs, not better DEQs but ones built to the 1919 specification? Let me hasten to add that elsewhere in the Admiralty, notably at the Signal School in Portmouth, there was enormously better expertise, but this did not always reach the rest of the Service.

Let us return to the test exercise. Submarine L27 left Fort Blockhouse at dawn the next morning and headed for the rendezvous in mid-Channel. Hetty Hyde, Hanbury and myself took off from Gosport in the Hudson a good deal later and arrived there about the appointed time. It was a dull day with a spanking wind and frequent rain squalls from the south-west. There was an angry sea and from 3000 feet we could scarcely see the surface, but we searched around by eye and by radar. Nothing showed and we began to think that this was going to be one of those gigantic muddles in which two Services, each entirely competent in themselves, would make a complete shambles of a joint exercise. But after 20 or 30 minutes a submarine was seen very tentatively emerging from the heavy seas below. We turned towards it gaily signalling on our downward recognition light, the first of the agreed signals. There was no response from the submarine. In fact, it crash dived and once again we were left with an empty sea. The answer seemed obvious. This was not the L27 but a German submarine, of which there were many in the Channel at that time. We had no depth charges aboard, but we had front guns. We waited for it to emerge once more and it did so about half an hour later.

Once again, we gave the identification signal on our downward recognition light, after which the proper response from the submarine was a grenade of the right colour combination. This did not appear. Playing cat and mouse, we carefully refrained from making any kind of run at the submarine—and waited for a response. Sure enough it came. As a fall-back, we had agreed that if there was a foul-up in the signals we would communicate by Aldis lamp. Up came morse signals from the submarine by Aldis lamp. Unfortunately they were so fast that nobody in the Hudson could read them. We had a hurried consultation and decided that, in spite of the absence of a grenade, this was almost certainly our submarine. We gave our own emergency signal for the start of the exercise—two greens from a Verey pistol—and hoped for the best. The submarine signalled back slowly enough for us to read, indicating that it was starting the exercise and its first course would be due east.

From then on the exercise went smoothly. There was very little of the submarine showing and the maximum range of detection was very small, but we completed several hours of runs, broadside and end-on to the submarine. After three or four hours exhausting work, we called it off and quit for the day. What with analysing the results and generally sorting out what had happened, we spent a busy afternoon, after which we foregathered at the submarine base to compare notes and make plans for the next day. It was late that night before the submarine was back at base and it was only then that we found out what had happened.

Submarine L27 had arrived at the rendezvous well before time and had cruised about on the surface waiting for us to appear. Sure enough, an aircraft with twin tails—a characteristic of the Hudson—had shown up and L27 fired a grenade to signify its presence. The aircraft responded by diving on them and letting go a stick of depth bombs. It was a Ju88 looking for just that kind of target. They dived and stayed down for 20 to 30 minutes. They came up very gingerly and took another look around. Visibility was not good, but there was another aircraft in the distance which they could not identify. The aircraft did not seem to be aware of their presence so they fired another grenade to attract attention. The aircraft swung round and did a run at them with 20 mm cannon blazing. It was an RAF Spitfire, intent on collecting a submarine if it possibly could. Submarine L27 dived again and waited a further 20 minutes.

When it came up a third time, we were the aircraft around and as we ran in, they fired their third granade. It turned out to be a dud and they had no more. There was nothing for it but to dive and have another long wait. They came up for a fourth time. This time, we had enough sense not to make a run at them; satisfied that we had no hostile intent, they started signalling by Aldis lamp. These were the signals which we could not read. We then fired the emergency 'start' signal and the exercise was under way.

It transpired that not once during the whole exercise had they seen our downward recognition signals—a light under the aircraft with which one could signal by morse code to a ship or vessel below. We went back to the aircraft at Gosport to check on this. The answer was soon obvious. The key for signalling morse code was alongside the instrument panel. Hidden underneath it was another small switch which decided whether you were signalling 'up' or 'down'. We had been signalling gaily in the *upward* direction. Until that moment we did not even know the switch existed!

I have described this at length to illustrate a problem which was endemic in military services the world over at that time. They each went their own way, developing equipment of one kind or another, completely unaware that a need would soon arise for them to communicate with each other. I was involved in a very similar incident off the east coast of America a couple of years later. A US Air Corps aircraft spotted a German submarine on the surface off the Atlantic coast but had no means of attacking it. The crew could see a US Navy destroyer within striking distance of the submarine, but they had no way of telling the destroyer what was going on. I am told that things are better these days, but I sometimes wonder.

The results of that first day at Gosport were not encouraging. From a height of 3000 feet the maximum range of detection of the submarine broadside-on was only about 3 miles, and it was even less in the end-on position. With a search area of literally hundreds of thousands of square miles to cover, that looked a hopelessly inadequate figure. However, Hyde and Hanbury went out next day and the results were better. The sea had subsided and more of the submarine's hull was showing. From a height of 5000 or 6000 feet, the maximum range rose to an encouraging 5 or 6 miles. This was in the dead ahead position and we knew that sideways-looking radar would improve on this by almost a factor of two. The Admiralty and Coastal Command concluded that this was better than nothing and, to their everlasting credit, pressed on with installations, both in Coastal Command aircraft and in aircraft of the Fleet Air Arm. This was the first step in the fight against the submarines, a battle which was to be fought with an ever increasing tempo for the remainder of the war.

When the Hudson finally called it a day and headed back to Gosport, they found it was closed by low cloud. They headed north for clearer weather and finally put down at an airfield in Oxfordshire. By a curious coincidence, the duty officer who greeted them at this field was Johnny Herd, son of Jimmy Herd who had been Watson Watt's deputy at the Radio Research Station, Slough.

The Battle of the Atlantic

As we have described, the first aircraft fitted with ASV Mark I were used mostly against surface ships and on escort duties. Only gradually were they deployed against submarines, and it was not until late in 1940 that the attack on submarines became the principal concern of

Coastal Command aircraft. At first, the hunt was concentrated on the north-west approaches and the Bay of Biscay, but gradually it spread deeper into the Atlantic as longer-range aircraft came into use. Submarine searches were a difficult and tedious job, and long hours of patrol led to few, if any, sightings. It was sometimes difficult to persuade crews that their very presence in the area would deter the submarines. This was a lesson which had been learned in World War I, but had been largely forgotten. In that war, unarmed training aircraft were used to make daylight patrols in coastal waters around Britain. When an enemy submarine spotted such an aircraft, it could not afford to wait to see if it was armed with bombs; it had to submerge at once and stay submerged until it thought the danger had passed. This was recognised as an important means of discouraging submarines from attacking shipping around Britain and it was used again along the Atlantic coast of the USA in World War II.

Prior to the introduction of radar, submarines made a practice of travelling on the surface at night. In winter, they could often remain on the surface for twelve or fourteen hours at a time with little possibility of being spotted by surface patrols. Their speed on the surface was 15 knots, as against 5 knots submerged, so submarines in transit preferred to make long passages at night, only submerging in daylight or when absolutely necessary. The new factor introduced by radar was that aircraft could now spot them on the surface in bad weather and at night. The aircraft might arrive at any time with no warning of approach and the submarines would be forced to submerge. This alone would seriously limit their mobility and the distance to which they could penetrate into the Atlantic. The task of the aircraft, therefore, was simply to hunt the submarines at night and keep them submerged as much as possible.

The first reports of submarine sightings from Coastal Command came in May 1940, the range of detection being four to six miles, as in our trials of the previous December. However, actual radar attacks were few and far between and it was not until 19 November 1940 that a Sunderland reported the first such attack on a submarine in the Bay of Biscay; unfortunately, the depth bombs failed to explode. Although these attacks may have frightened a few submarines, there was continual trouble with depth bombs and very little damage was done. Throughout 1940, shipping losses continued to increase at an alarming rate, and air operations against submarines that year must be judged a failure.

The horror of the situation is shown in a graphic way in the curve of shipping versus submarine losses given in figure 6.1. This shows the ratio of the number of Allied vessels sunk to the number of German submarines lost from all causes, computed at three-monthly intervals.

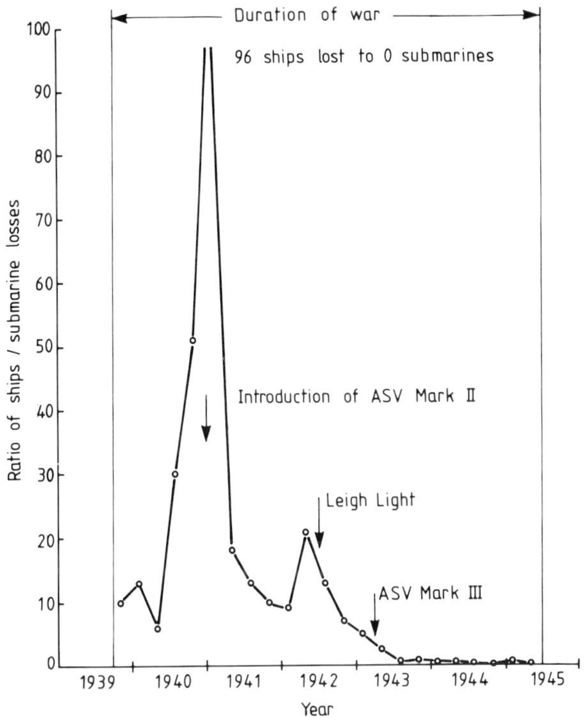

Figure 6.1 The ratio of Allied ships sunk to the number of German submarines lost, plotted at three-monthly intervals for the duration of the war. The dates on which ASV Mark II, the Leigh Light and ASV Mark III were introduced are as shown.

During the first nine months of the war, the ratio was already at the frightening level of ten ships sunk for every submarine lost. The loss ratio mounted rapidly through 1940, until during the three months December, January and February of 1940–1, no less than 96 Allied vessels were lost and not a single enemy submarine. If that level had been maintained, the war would have been quickly over.

The failure of the attacks during 1940 can be put down to many factors, but in simple terms it was due to the need for a very protracted training period, during which crews of Coastal Command and the Fleet Air Arm learned how best to use their new equipment. Their problems were compounded by the poor serviceability of ASV Mark I, the lack of test equipment and the total absence of training facilities. It remains true to this day that any military service, when faced with the introduction of a new and exotic equipment, requires meticulous training and a long and protracted practice period before they can make best use of it.

Figure 6.2 A Short Sunderland of RAAF 10 Squadron, fitted with long-range ASV aerials, at Pembroke Dock. (Courtesy of the Imperial War Museum.)

Figure 6.3 A Vickers Wellington with long-range ASV aerials over the Scottish lochs. (Courtesy of the Imperial War Museum.)

Figure 6.4 A Fairey Swordfish. This aircraft did sterling work throughout the war as a sea-search aircraft with the Fleet Air Arm. The ASV aerials are showing on the outer struts. (Courtesy of the Imperial War Museum.)

Figure 6.5 A rare example of a Boeing Flying Fortress fitted with ASV Mark II manufactured in the USA. (Courtesy of the Imperial War Museum.)

Towards the end of 1940, the much improved ASV Mark II came into operational use and was fitted to a wide variety of Coastal Command aircraft—Hudsons, Whitleys, Wellingtons, Sunderlands and Catalinas—many of which carried both forward- and sideways-looking aerials. These were followed by Swordfish, Albacores and still more Catalinas, operated by the Fleet Air Arm. Figures 6.2, 6.3, 6.4 and 6.5, show some typical aircraft installations of that period. By the summer of 1941, more than 100 ASV-equipped aircraft were being operated by Coastal Command. The Hudsons operated over the North Sea and the Whitleys over the north-western approaches, while Sunderlands and Wellingtons covered the Bay of Biscay and the nearer reaches of the Atlantic. As experience built up and crew training began to pay off, the number of attacks on submarines increased and a change came over the scene. The first authentic case of damage being inflicted came on 10 February 1941 and, although there were still no sinkings, attacks from aircraft now occurred often enough to scare the submarines and keep them submerged at night. This limited their movements to such an extent that, for the first time during the war, shipping losses showed a significant reduction. As shown in figure 6.1, there was a dramatic reduction in the number of ships lost and by the end of 1941 it was back to a level of ten Allied ships lost for every enemy submarine. This was still an unacceptable figure, but it was a far cry from the carnage of the winter of 1940–41.

While ship losses were coming down, the number of aircraft equipped with ASV radar was building up, and during the next twelve months the total number of aircraft engaged on anti-submarine duties reached 400, three-quarters of them equipped with ASV Mark II. The total number of German submarines at sea seldom exceeded 50 in any one month, so that for every submarine there were 8 or 10 aircraft searching for them over the Atlantic and the coastal approaches. During most of that year, radar-equipped aircraft were attacking submarines but not sinking them. However, the first successful night attack came on 30 November 1941 when a Whitley of 502 Squadron, equipped with ASV Mark II, attacked and sank submarine U206 in the Bay of Biscay. The tempo then increased rapidly. Just a few weeks later, Fleet Air Arm Swordfish aircraft of 812 Squadron operating from Gibraltar had a remarkable run of success against submarines trying to pass through the Straits into the Mediterranean. In quick succession they hit and severely damaged no less than five submarines at night, all of which were forced to return to their bases at Brest and St Nazaire for repairs. Then on 21 December, Sub-Lieutenant Wilkinson secured the first kill in this region when he attacked and sank U451. The number of successful attacks continued to increase, but at the beginning of 1942 there was a set-back. There were two reasons for this. Scare tactics could not succeed for ever. The German submarines decided that, although they were being attacked at night, they would stick it out and remain on the surface. As shown in figure 6.1, during the spring of 1942 the ratio of ship to submarine losses crept up again from 10 to 20. This was a shocking figure and there was a danger that it would mount even higher. A second reason for the increase was that America had now come into the war; the submarines shifted their attack to the eastern seaboard of the USA and, for a time, made a feast of relatively unprotected shipping moving along that coast.

It was at this point that, once again, a new technology came to the rescue. This time it was in the form of the Leigh Light, developed more or less as a personal venture by Squadron Leader Leigh—an administrative officer at Coastal Command. He had been a pilot in World War I when he had participated in submarine searches in the Mediterranean. In conversation with Squadron Leader Lugg, who was still the officer principally concerned with supervising the fitting of radar to Coastal Command aircraft, Leigh recognised the problem. Radar-equipped aircraft could find submarines at night, but during the final run-in the pilot and the bomb aimer could not see the target. Squadron Leader Leigh proposed mounting a searchlight in the search aircraft which would then make a low-level run on a surfaced submarine, illuminate it and drop depth charges with nearly the same accuracy as in daylight. This equipment took some time to develop but by June of 1942 it came into increasing use in Coastal Command aircraft.

The combination of the Leigh Light and ASV radar was extraordinarily effective and had such a demoralising effect on the German submarine crews that they actually switched their routine. The night attacks became so severe that they took the unprecedented step of remaining submerged at night and coming to the surface by day to recharge their batteries. They did this on the grounds that there was a reasonable chance of seeing an aircraft approaching in daylight, but none at all at night. From the point of view of the Allies, this had a two-fold effect. In addition to the dramatic increase in night sinkings, there was now a significant increase in the number of daylight sightings. The result was inevitable—the Navy intensified their daylight patrols and the number of submarine losses mounted even higher.

Finally, the submarines came back with a counter-measure in the form of a listening receiver which told them when a radar-equipped aircraft was approaching. We now know that this came about because the German High Command became aware of the British use of 200 Mc/s ASV radar from another source. Rommel had captured a serviceable Wellington fitted with Mark II ASV in Tunisia and used it with great success for reconnaissance against our ships in the Mediterranean. When the German High Command heard about this, they recalled the Wellington to Berlin for test. They were impressed by its performance and started work on a German version of ASV radar. Of more significance to the submarine war, the Germans designed a listening receiver—'Metox'—which enabled submarine crews to pick up the radar signals from an approaching aircraft and submerge before the aircraft radar could detect the submarine. For the first time, there were reports of disappearing radar contacts, which indicated without doubt that the submarines were detecting the approach of our radar-equipped aircraft and diving before an attack could be made.

This occurred in the latter part of 1942; there was a serious danger that the submarines would make a come-back and that shipping losses would increase once more. Drastic measures were called for. Luckily, for reasons we shall describe in later chapters, the Allies were ready with the answer—a sea-search radar of superior performance operating on the extremely short wavelength of 10 cm. It was called ASV Mark III and the first Wellingtons equipped with Mark III and the Leigh Light went into operation in March 1943; it completely confused the German submarine crews. Their receiver was unable to give them warning of the approach of an aircraft with a centimetre-wave radar and once again they were being illuminated and attacked on the surface at night, even more accurately than during the previous winter.

This was the beginning of the end for the German Submarine Service. As shown in figure 6.1, from the summer of 1943 until the end of the war the ratio of ships lost to submarine losses was reduced to the remarkably low figure of 0.8. That is, for every submarine sunk, on average less than one Allied ship went down. Not all the authorities are agreed on the figures for losses sustained by the two sides, but the figures given in figure 6.1 are derived from official German Naval sources and given by Harald Busch in his book *U-Boats at War* (1955).

However, incredible as it may seem, centimetre-wave ASV radar was not introduced without a fight of a different kind back in Whitehall. This was a political, not a technical, battle and as on earlier occasions, it was a case of Professor Lindemann (now Lord Cherwell) going against the judgement of Sir Henry Tizard. Tizard, strongly backed by Blackett, pointed out the vital importance of keeping the sea lanes open. He argued that the first use of the new centimetre-wave radars should be against submarines. Cherwell wanted them used as a bombing aid against Germany. The argument was long and intense, with the Admiralty firmly on the side of Tizard and Blackett. Finally, a compromise was reached and a few sets were made available for use over the Atlantic. In describing this confrontation, Captain

S W Roskill in his book *The War at Sea* says that the last stage of the Atlantic battle:

> . . . was grievously handicapped by a shortage of suitable aircraft. For what it is worth, this writer's view is that in the early spring of 1943, we had a narrow escape from defeat in the Atlantic. Had we suffered that defeat, history would have judged that the main cause had been the lack of just two more squadrons of radar equipped very long range aircraft for escort and search duties.

Such are the narrow limits between success and failure in modern warfare.

By concentrating this account on the radar battle, wea are in danger of presenting an oversimplifed picture of the Atlantic battle. Many differenet methods of detecting submarines were used—Asdic, magnetic detection, radar alone and radar in conjunction with the Leigh Light, to name only a few. In addition, submarines were attacked by a wide variety of weapons—depth bombs, cannon, rockets, mines and, ultimately, homing torpedos—all of which played a vital part in their final defeat. For a complete story, it is important to examine the total losses of ships and submarines as well as the ratios shown in figure 6.1. This is well beyond the scope of the present work and the reader is referred to Alfred Price's book, *Aircraft Versus the Submarine* (1973), for a more complete story.

The Battle of the Atlantic was a momentous struggle, with terrible losses on both sides, Throughout the war, a total of 2603 Allied merchant ships were sunk, carrying thousands of invaluable men and millions of tons of food and raw materials. Against this, out of nearly 1200 German submarines which put to sea, a total of 783, or 65 per cent, were irretrievably lost. About half of these were sunk by aircraft operating alone or in conjunction with Naval vessels; the other half fell to Naval vessels. Forty thousand men were recruited into the German Submarine Service; no less than 28000 perished and a further 5000 were taken prisoner.

The slaughter continued right up to the end of the war. During the single month of April 1945, 57 German submarines were destroyed from all causes. This was at a rate of almost 2 a day and, with 30 or 40 crew members per submarine, it hardly bears thinking about in humanitarian terms. Most of these figures relate to the Atlantic Ocean and adjoining waters. A few German submarines penetrated as far as the Indian and Pacific Oceans, where they did substantial damage. Out of the 41 which ventured that far, only 4 returned to their homeland.

With the war at sea going completely against him and with the Reich tumbling about his ears, Hitler himself is quoted as saying:

> It was one single weapon . . . which defeated the German Submarine Service.

That weapon was ASV radar in its many forms. Right at the end of the war, Admiral Doenitz put it more explicitly. He said:

> The enemy has deprived the U-Boat of its essential feature—namely the element of surprise—by means of radar. With these methods . . . he has conquered the U-Boat menace. The scientists who have created radar have been called the saviours of their country.

7

Introduction of Air-Interception Radar into Service

We left the air-interception story at the point where the first rushed installations had been made in Blenheims at Martlesham Heath, just before the outbreak of war. A batch of six were despatched to 25 Squadron at Northolt in August 1939 and the remainder completed at Perth. The main lesson learned from the Northolt exercise was that a much more accurate ground control system than that currently provided was needed before night interception was likely to be successful. Also, that the task of carrying out a night interception was very much more exacting than had previously been realised and that an exceptional degree of skill and training of pilots and operators was required. Partly for this reason, 25 Squadron was moved to Martlesham so as to be near Bawdsey, where night inteception experiments under direct control from a chain station could more conveniently be carried out. In effect, the Squadron became a training centre for pilots and AI operators—something which had simply not existed up to that point.

The rush to get AI radar into service may have been justified, but it led to some awesome problems. The pilots had no idea what to expect from the new equipment and they had no prior training in the very special technique necessary to pull off a night interception. There was a natural tendency on their part to think in terms of day fighting, with aircraft diving and swooping around the sky in all directions. A night pursuit is something entirely different. It calls for a very gradual and stealthy approach. First, the pilot must creep slowly up on the target, rather like a hunter stalking his prey. He must then close gradually to minimum range and place himself in exactly the right position behind the target before actually opening fire.

Apart from this, there were no trained radar operators as such. The first operators were recruited and trained as rear gunners and, with the best will in the world, very few of them had the temperament necessary to handle exotic electronic equipment. None of them had previous experience of handling the delicate focusing, gain and tuning contols needed in a radar set. The problem of finding skilled radar operators was not solved until someone at Air Force HQ had a brain-wave and revoked the previously inflexible rule that individuals wearing glasses could not act as aircrew. This allowed an intake of schoolteachers to become radar operators and the problem disappeared almost overnight. There were also serious maintenance problems. The equipment itself was not well engineered. At first, it simply could not stand up to the vibration and the altitude problems found in aircraft. In addition, the average Air Force radio mechanic was not trained to cope with unusual components like acorn valves and cathode-ray tubes. Needless to say, there were no technical manuals and no test equipment to help them. Here again, it was only after there was a special intake of ham radio operators from civilian life and after training schools had been set up that the problem was ultimately solved.

The second squadron to be equipped from St Athan was a unit of the Auxiliary Air Force, the City of London 600 Squadron. This was an elite squadron made up of prominent City bankers and stockbrokers. Like 25 Squadron, they were also based at Northolt and, in the absence of proper headquarters, were parked around the perimeter in caravans. There were no air attacks of any consequence and everything went along very much as it had in peacetime. If the truth be told, many of them had a private telephone connection from their caravans to the office in the City and in between flying duties they continued to do a little business on the side—strictly picnic fashion and in the best amateur tradition. The atmosphere contrasted sharply with that at St Athan, where we wondered how we could grab another hour or two to attend to the next pressing task.

The Commanding Officer of 600 Squadron was Lord Carlow, who had a considerable pull at Fighter Command and ran his Squadron like a private air force. He was among the first to realise that the Blenheim in its basic form was hopelessly underarmed—it had one .303 firing forward from the wing and one, or sometimes two, .303s in a rear turret, which was useless for night fighting. He therefore equipped his aircraft with an additional battery of four .303s in a tray buttoned under the belly. Finding the Air Ministry reluctant to make this modification, he arranged for it to be done at his own expense. He also did much to see that his pilots were protected from enemy fire by the installation of armour plate behind and underneath the pilot's seat. It reminded me of the days at Bawdsey when we used to drive into Ipswich or send someone to London to buy specially needed components with our own money. In some respects it was still a very peculiar war.

Lord Carlow visited us several times at St Athan to see how the work on his aircraft was coming along. He impressed me immensely as a Commanding Officer who took a real interest in the problem of night fighting and was prepared to put in a great deal of effort to ensure that it would be successful. He already knew the broad outlines of what was required from his colleagues in 25 Squadron and he made an all-out effort to make things work. We had already delivered a number of Blenheims to 600 Squadron when, suddenly, we heard no more. It was January or February 1940 when the Russians sided with the Germans and started to attack the totally innocent Finns. Without warning, Lord Carlow was ordered to strip his aircraft of their AI equipment and transfer to Finland, together with his ground crews. A total of nine aircraft were involved and this meant that all our preparations and the instruction his crews had received in night fighting were lost in a few hours. Even more distressing was the fact that his Blenheims, which landed on a frozen lake in Finland, proved totally unsuitable for operation in the snow and ice of a Finnish winter and suffered a high attrition rate. They lost most of their top flight crews, including Lord Carlow himself. I have tried to track down the record, but have been quite unable to find any official account of the expedition or to find out if any of that gallant band of men ever returned to Great Britain.

By this time, the AI equipment had moved from Mark I to Mark II, and was then materially improved by the addition of micropups from GEC which brought the transmitter power up to something over five kilowatts. This was AI Mark III, which gave very healthy signals on a target aircraft at a range of three to four miles. We continued to send aircraft equipped with AI Mark III out of St Athan at a rate which had now reached almost one a day.

In the meantime, 600 Squadron reformed with still more Blenheims, and moved to Manston in Kent—the most forward aerodrome in Britain. The personnel of 600 Squadron were entirely new and had to be trained from scratch. Not a single member of Lord Carlow's Squadron remained. I started the training process again and spent a good deal of my time at Manston in April 1940, giving them instruction in the use of the equipment and demonstrating interception techniques. The new Commanding Officer of 600 Squadron was Squadron Leader Anderson and his deputy was Flight Lieutenant

Davison. During one flight with Anderson as pilot, I can remember a frightening moment. We were above cloud at 14000 or 15000 feet, doing a routine test interception on another squadron aircraft when I suddenly realised from the radar screen that there was another aircraft within range and closing fast. I did not like the look of it; I whipped my head out from under the black cloth and looked around. There was a Hurricane racing in from the starboard side, clearly intent on getting us in his sights. I was scared to death at the sight of those eight guns and yelled to Anderson to dive for the cloud cover which was just below us. We made it just in time and promptly abandoned the exercise.

It turned out that the Germans had reconditioned several Blenheims which had crashed over Germany and were flying them up the Thames estuary for reconnaissance in the London area. There was an Air Ministry order which said that any Blenheim found flying over the Thames estuary was to be shot down on sight. There was thick cloud and we had no way of knowing whether we were over the estuary or not, but it was a salutary experience. More than one British aircraft was lost in this way about that time.

A few days later, on the morning of 10 May 1940, for the second time in its history, 600 Squadron was ordered to strip out its AI equipment and attack the Germans over Rotterdam. They could have been the only squadron so deployed—six ill-equipped aircraft against the might of the German Air Force. Needless to say, they were practically wiped out. Of the six aircraft, Flight Lieutenant Davison was the only one to return, with one kill to his credit. The other five were shot down; a few of the aircrew were taken prisoner, but the majority were never accounted for. Apart from Flight Lieutenant Davison, one other pilot and his radar operator returned to Manston under quite remarkable circumstances. Their aircraft was shot up and badly damaged, but they managed to make a belly landing near the Dutch coast. They hid overnight in the dykes and next morning came across a 15 foot boat on the beach, which they commandeered and sailed back to the North Foreland. They landed at Margate, hired a taxi and drove back to Station Headquarters at Manston. They had no money, so they asked the driver to wait while they raised the fare from the Commanding Officer!

So ended the second attempt to equip 600 Squadron with night fighting AI equipment and it took a long time to recover from this set-back. Flight Lieutenant Davison was given an immediate DFC and posted to another squadron. Until it was reformed for the third time with new aircraft and new AI equipment, 600 Squadron simply ceased to exist.

Fighter Command could certainly not be blamed for this. For many months, Stuffy Dowding had been under intense pressure from the politicians to transfer Spitfires and Hurricanes to France as a defence against the coming German attack. Without the back-up of a chain air-warning system, this was something he stoutly resisted. It very nearly cost him his job, but events proved him to be right. The Spitfires and Hurricanes remained in Britain and shortly scored a resounding victory over the German Air Force. Presumably, Stuffy was not able to resist the political pressure to deploy at least some of his aircraft during the Finnish crisis, or again when the Gemans pushed through the Low Country. In its ability to fight the coming night battle over Britain, the country paid a considerable price. When some devastating attacks developed against British cities in the latter part of 1940, the events I have just recounted undoubtedly contributed to the lack of night fighters to counter that threat.

It also confused the issue. As 1940 drew to a close, night fighters were simply not shooting down German aircraft at night, and the very concept of using radar-equipped fighters for night interception came under criticism. The source of this criticism could not be clearly defined, but it was undoubtedly fuelled by the nagging refrain from Lord Cherwell, who lost no opportunity of throwing spanners into the radar works. It gave rise to a whole host of competing ideas. One of these was the so-called Turbinlite, a searchlight carried in the nose of a fighter to illuminate its target. Another was bombing enemy aircraft from above; many other schemes were proposed, none of which merit a detailed description. They were seldom thought through in explicit terms and not one of them developed into a useful method of defence against night air attack.

The next squadron to be equipped with AI radar was 604 Squadron, another City of London Auxiliary Squadron. Our commitments were so heavy that there was no way we could give them, or the other squadrons formed about that time, any direct help with their equipment. They were virtually self-taught, having to find out for themselves just how to make best use of the mysterious black boxes in their aircraft. After a long and frustrating training period, 604 Squadron was destined to have the best record of any night fighter squadron during the war. They were led by Squadron Leader John Cunningham, the most distinguished night fighter pilot of the war, with 20 night kills to his credit. Much of this story will be found in the excellent book *Night Fighter* (1957) by Jim Rawnsley and Bob Wright, who were themselves night fighters of considerable merit.

We did not see a great deal of 'Blood-orange' Smith at Perth and St Athan. He was still a member of 'D' flight but went back to Martlesham to take charge of calibration flying for the east coast chain. He took several radar-equipped Blenheims with him and, in between calibration flights, he tried valiantly to be the first RAF flier to score a kill at night with radar, under direction from Bawdsey. Unfortunately he was not successful at this, but being Smith, he soon distinguished himself in an unexpected way. Like many others, his blood boiled when the Germans went through the Low Country and attacked Rotterdam. He decided to fight the war his own way and arranged a radar calibration flight which took him right over the top of that city. There he found literally scores of German aircraft, all too busy to concern themselves with a lone British intruder. He went down into the fray, picked out an He 111 and shot it down. He then made the mistake of picking a second one. He put the rear gunner out of action from long range and as he had practically run out of ammunition, went in close to finish the action. As Smith described it later: 'You see, it was like this. I put the rear gunner out, but the front gunner was a bad type. He climbed back through the fuselage, manned the rear gun and opened fire.' Smith's radar operator was wounded, his port engine was put out of action and Smith himself suffered a broken forearm, a broken collar bone and was losing blood from wounds across his chest.

No one knows how he made it home, but he did. To add to his problems, the port propeller would not feather and, 15 miles from Martlesham, the port engine caught fire. He came into the field with the engine well alight and a huge plume of black smoke behind him. When he touched down, his radar operator was already half way out of the back cockpit and fell off in the middle of the field.

If they overran the field, the Blenheims of that period had a nasty habit of disintegrating in a ball of fire. Smith realised he was going to overshoot and abandoned ship through the top hatch. Moments later, the Blenheim hit an obstruction at the far end of the field and exploded in a blinding flash. Those who saw it swear the bits went a thousand feet into the air. Some pieces of molten aluminium were on display as hallowed relics at 'D' flight for a long time afterwards. Smith himself made a miraculous recovery. In a little over two months, he was on light duties and shortly afterwards he was flying again. The Commanding Officer at Martlesham recommended him for an AFC, citing his bravery in bringing back a wounded crew member. The awards committee in London altered this, pointing out that what he had done was in the face of the enemy and therefore deserved the DFC, which was promptly awarded.

About this time, in between many other commitments, I had been amusing myself working on an elementary theory of air interception. The easiest case to think about was the classic stern chase, in which the pursuing aircraft simply pointed directly at the target and kept going until he reached a position directly behind it and opened fire.

Several things came out of the study. First, and terribly obvious, was that the pursuing aircraft must have a speed advantage over the target. If the fighter were no faster than the target, it would never catch it. If on the other hand it went too fast, it would overshoot. The analysis showed that there was an optimum overtaking speed of about 20 or 25 per cent. This simply meant that if the target was flying at 250 mph, as well it might, the optimum speed of the pursuing aircraft was 300 mph. I gave briefings at Fighter Command which emphasised this point. I can remember one or two officers squirmed on hearing the figures and I wondered why. I was still influenced by the propaganda which said that the Blenheim was one of the fastest medium bombers in the air and would catch anything the Germans cared to put against it.

A second conclusion from the study was, that in order to have a reasonable chance of completing an interception, the night fighter must be placed within a cone of about 40 or 50 degrees behind the raider and heading on a track not more than 30 degrees different from the target. If it approached from far out on the beam, that is to say at right angles to the enemy's track, he had little chance of completing an interception. This presented a formidable task to the ground control system as it existed at that time and was not solved until the appearance of special GCI equipment towards the end of 1940.

These were simple ideas which came from our experience in the air and from some very elementary calculations of the path likely to be followed by a radar-equipped aircraft in pursuit of an enemy aircraft. We had lived with this problem for a couple of years and knew it like the veins on the back of our hands. The problem was how to communicate these skills to the new generation of night fighters. Clearly, what was required was an intermediate step between the experimental flying we had done at Martlesham Heath and actual squadron use; that is, a unit which would establish procedures acceptable to the RAF, covering everything from routine ground tests of the radar equipment, to daily flight checks, to checks of the interception process itself. Tizard and Watson Watt both saw the need and as a result a Fighter Interception Unit (FIU) was formed to carry this responsibility.

Fighter Interception Unit

The Fighter Interception Unit was set up at RAF Tangmere under the command of Squadron Leader Peter Chamberlain and equipped with five Blemheims from St Athan. I was specially asked to be present for the first few weeks and was at Tangmere to meet the aircrew when they arrived on 20 April 1940. I gave them an intensive series of talks on the whole radar story, from the design of the equipment to the elements of its use in the air. Radar was still a super-secret and few Air Force officers knew what the mysterious black boxes were all about. Even in 1940, there were very few people in the whole of Britain who knew both the details of the equipment and the intricate procedures of night interception. These were Keith Wood, Chalky White, Hanbury Brown and myself—all civilians who had a fair amount of technical expertise but were lacking in proper Service experience. We demonstrated our version of the interception procedures and very rapidly the FIU came to understand the intricacies of the process, probably better than we did ourselves. At the end of April I had to break away from Tangmere for a few days but returned again on 1 May for two more weeks. This time I took Chalky White with me and, as he was going to remain there for some time, we drove up to Tangmere from St Athan in Chalky's car, a red sports model of which he was justly proud.

I had known Peter Chamberlain since he was Chief Signals Officer of Coastal Command at Lee-on-Solent in 1938. He was a very dynamic individual who most decidedly led from the front. He was always the first to try any new procedure and had a clear and incisive mind. He never seemed to sleep. He did a full quota of daily test flights and, although the Unit was not expected to take on the Germans, he insisted on charging off into the night at the least sign of enemy activity. His number two was Flight Sergeant 'Jumbo' Ashfield, one of that extraordinarily talented band of pilots which the Air Force is always able to produce when needed. To me, he was one of the most likeable and most competent pilots I have ever come across. He could also put the fear of God into relatively staid and conservative people like myself. I am most comfortable flying straight and level in an ambulatory aircraft like the Anson. Jumbo Ashfield soon fixed that. The Blenheim was far from being an aerobatic aircraft, but he put those FIU machines into attitudes which would have frightened the life out of their designers and certainly frightened the daylights out of me. He was partial to low flying and I still have recurrent dreams when I imagine I am flying at high speed in and out of trees and under power lines–with Jumbo Ashfield at the controls, a broad grin on his face.

The routine of trial interceptions by day and excursions against the enemy at night continued and was finally rewarded when Jumbo Ashfield recorded the first kill against a night bomber. About midnight on the night of 22–23 July 1940, he was on patrol just off Bognor when he was directed to a raider by the Poling chain station. He soon made contact with his AI radar, closed to minimum range and found himself beautifully placed to attack an unsuspecting Do 17. He opened fire from very short range; the Dornier must still have been carrying its bombs, because it blew up in his face, showered his aircraft with debris and covered his windscreen with oil. The force of the explosion turned the Blenheim on its back and Jumbo was hard put to recover before hitting the sea. This was the very first enemy aircraft to be shot down at night by a radar-equipped night fighter, but the victim was not seen to fall into the sea so there was a momentary doubt about its identity. However, all doubts were removed when the mangled remains of the Dornier were recovered from shallow water off Bognor a few days later. For a brief period, we thought our problems were over, but the number of successful attacks remained disappointingly small. In the next three months, that is by the end of October 1940, only six more enemy aircraft were shot down by Blenheims carrying AI Mark III.

In the midst of activities at Tangmere we were not without some interesting diversions. France had been told about British radar developments before the outbreak of war and some desperate efforts were being made to shore up their activities. The first British Mobile and Gun-laying radars had already been sent to France, but up to that time the French had no airborne radar and no radar identification equipment to operate with the ground radars. Although it was not my direct responsibility, I was asked to keep a friendly eye on the installation of an IFF set into a French aircraft which would shortly arrive at Tangmere; it was the first such equipment to be installed in an aircraft of the French Air Force. The aircraft duly arrived—a Potez 63. II. On its arrival, it put up a dazzling display of aerobatics over the field at Tangmere. It did upward rolls, loops, the whole act for the entertainment of the Station staff who had come out to see the show. It finally came in for a dignified and sedate landing. The wheels came down, it made a beautiful touch-down but after running a few hundred yards, ground looped and dug a wing tip into the dirt. Nothing could be more embarrassing. The crew tumbled out of the aircraft, peered under the wings and said '*Toujours la même musique.* It always does that.' One of the tyres was flat.

Not much damage had been done but worse was to follow. A whole gaggle of Air Commodores, Wing Commanders and their aides had assembled to welcome this, the first French aircraft to arrive in Britain on a technical mission. First out of the aircraft was the pilot. He had no uniform of any kind. He was the nearest approximation to a Paris taxi driver one could imagine. He was dressed in a leather jerkin, breeches and riding boots, capped by a beret. Similarly, the radio operator was as near as one could get to a Breton fisherman, in black jersey and whipcord trousers. The rear gunner wore a navy blue uniform jacket which may have given him Warrant Officer status, but his trousers were as nondescript as those of his companions. For the first appearance of a French Air Force machine in Great Britain it was a pretty poor show.

However, the crew were quite relaxed about it and quickly made friends around the station. An IFF was fitted to the Potez, it was flight-tested against the Ventnor chain station and pronounced satisfactory. The French Air Force had its first piece of radar equipment in an aircraft and the time came to say farewell. That slow puncture in the Potez had never been fixed and on each landing there had been some ominous swings, but no actual ground loop as on the first landing. The day of departure came and appropriate sentiments were expressed. We were genuinely sorry to see them go—the Paris taxi driver, the Breton rear gunner and the Warrant Officer radio operator. The tyre was pumped up for the last time, they ran down the runway and gave another dazzling performance. Not one, but two upward rolls off the deck, followed by some sparkling aerobatics. There was a pause while they prepared to do a loop at minimum altitude. We sensed that this was their last fling before departing. They dived, built up speed and pulled up until they were pointing straight up into the sky—and the port propellor came off. They recovered as best they could and made a final landing. Crestfallen, they helped push the Potez into a large hangar opposite the Officers' Mess, pending repairs.

Meanwhile, the situation back in France was becoming desperate and the crew were recalled without their aircraft. I returned to Tangmere a few weeks later and found that the hangar in which the Potez had been stored had been wiped out by a direct hit from a 1000 pound bomb. The Potez was reduced to a pancake not more than a few inches thick, flattened under those great hangar doors. What I did not know until many years later was that under those same doors was Chalky White's beautiful red car.

Difficulty of Air Interception in Blenheims

In a previous section I mentioned the need for a 25 per cent excess of speed required by a night fighter in pursuit of its target and our forebodings about the ability of the Blenheim in this direction. The problem was brought home to us vividly one night when, somewhere over Brighton, Jumbo Ashfield, with Chalky White as his radar operator, got onto the tail of a bomber returning from a sporadic raid on the south coast. They crept up in approved style, but when they came close to visual range the enemy aircraft started to pull away.

Jumbo went to maximum boost but nothing he could do would shorten the range. Occasionally, the target was almost lost off the end of the range scale, but somehow they managed to hang on. They were so intent on the chase that, as had happened many times before, they failed to keep track of their position. They were pulled up short by no less a person than Air Chief-Marshal Stuffy Dowding, the legendary Commander-in-Chief of Fighter Command himself.

It was well past midnight and before retiring to bed at Fighter Command Headquarters at Stanmore, Stuffy had gone down to the Filter Room to see if there was any action. It happened to be a quiet night and there was a single pair of plots on the board. It was the lone German raider being chased by Jumbo Ashfield and Chalky White. Stuffy asked the officer in charge which squadron this was and who were the crew. Rather nonchalantly, he was told that it was an FIU aircraft carrying the latest brand of AI radar with Chalky White on board. If the Commander-in-Chief cared to wait a moment, perhaps he would see the enemy shot down. It turned out that both aircraft were already over the Belgian border and in a few minutes would be over German territory. 'Call them back at once' cried Stuffy angrily. He was a bluff old character, willing to stake a great deal, but he was not going to take the chance of losing a precious radar-equipped aircraft over German soil and having a civilian shot at dawn for violating the Hague Convention.

So, that particular chase was called off, but it give a clear indication of where the difficulty lay. We were still certain that airborne radar was the solution to the night interception problem and highly optimistic about the performance of our equipment, but if the aircraft carrying it was not fast enough to catch the opposition, what hope did we have? The unpleasant truth was that if a Blenheim was going to catch a German aircraft over Britain, that aircraft had to be dawdling. There was still an understandable tendency to put the blame for these failures on the air-interception radar itself, or even on the whole concept of a radar-directed night defence, and we often had to argue the case in favour of radar. But we now knew that the problem was simply one of speed of the night fighter and confirmation soon came from the squadrons. There was a spate of reports of AI contacts being made after which the target disappeared off the end of the trace—the so-called disappearing contacts. It became very clear that the Blenheim was underpowered and simply not fast enough to catch the German raiders. Even in the rare event of a Blenheim coming within firing range, its fire-power was pathetically small. In the quaint official language of the time, a typical report of yet another unsuccessful night interception would conclude 'With respect, Sir, but the Blenheim was not fast enough.'

This was a sad and frustrating period when, night after night, the Germans made grievous attacks against London and cities elsewhere in Britain, with terrible devastation and much loss of life. The unpleasant truth was that the night defences of the country did not have the equipment or the trained men to cope with the attacks. It was a similar problem to that already being experienced with ASV radar against submarines. Time was needed for the equipment to be incorporated into the Air Force system and for its operation to be completely understood. Above all, there was the need for a long practice period during which the complex routine of an air interception was repeated, time and time again, until its execution was perfect in every respect. In basic terms, the whole operation must be so much part of the system that it becomes entirely automatic. Only when that stage is reached is it likely to be successful.

There were, in fact, a multitude of problems, all of which required time to resolve. There were difficulties due to inadequate airfield lighting and the almost complete absence of navigational aids. There were even problems in communication between the pilot and his radar observer due to the poor intercom system, and many other distractions of a similar kind. These are expounded in a very clear and philosophical fashion in the book *Cover of Darkness* by Group Captain Roderick Chisholm (1953). Watson Watt has very aptly said that this book should be subtitled 'Portrait of the Artist as a Night Fighter.' It gives a penetrating and sympathetic account of the problems met by the night fighters in those early days.

However, salvation was around the corner. One quiet afternoon at Tangmere when air tests were done and crews were resting in anticipation of another night of activity, Jumbo Ashfield said 'Let's go down to Bristol and look at the Beaufighter.' This was the very first occasion on which I had heard of the new aircraft—the one which was destined to turn the tide of the night battle. We hopped into a Puss Moth and headed for the Bristol aircraft factory. When we got within sight of the field Jumbo announced in his nonchalant way that now we had better be careful. The factory and its adjacent landing ground were entirely surrounded by a balloon barrage to protect them against a dive-bomber attack. There was only one way in —to approach from due north, do a right-angle turn when we were abeam the runway and land into the setting sun. To a medium or heavy bomber, those wires would not present too much of a problem, but for someone in a fragile Puss Moth, they did not look good at all.

We landed safely and were taken to a large hangar where we were treated to a heart-warming sight—the first Beaufighter being assembled on the shop floor. To this day, the Beaufighter is an aircraft which radiates an extraordinary aura of power—both engine power and fire-power. The internal layout seemed ideal for a radar-equipped aircraft and Jumbo and I looked in awe at the powerful Hercules engines and the four 20 mm cannon nestling in the belly. With four to six more .303s in the wings, this was indeed a formidable machine. We headed for home with much lighter hearts. A Beaufighter equipped with the latest AI radar clearly had the potential to make an enormous difference to the coming night battles.

Introduction of the Beaufighter and Ground Control Interception Radar to Night Fighting

The first Beaufighter, R2055, arrived at FIU on 12 August 1940 and was fitted with a Mark IV AI radar—the final version of that particular brand of AI radar. There was a momentary hitch when the aircraft was damaged during a daylight attack on Tangmere, after which FIU moved to temporary quarters at Shoreham. The first operational sortie was made by Jumbo Ashfield on 4–5 September but, unfortunately, after only two or three such flights, the aircraft was lost on the night of 12–13 September. It was thought to have gone down into the sea off Boulogne with Flight Lieutenant Ker-Ramsey as pilot.

It was subsequently found that Flight Lieutenant Ker-Ramsey was taken prisoner of war, which suggests that the aircraft was not greatly damaged. Whether it was salvaged by the Germans, who thus found the secret of the air-interception radar, remains a mystery to this day. However, it is on record that very shortly afterwards General Martini of German Air Intelligence asked the Telefunken Company whether it was possible to build an airborne radar to assist the German night fighters operating in Holland. Construction of a German version of airborne radar began early in 1941 and the design had many features which were similar to the British model, including the display system which was identical.

The lost Beaufighter was replaced by R2059, and it was in this aircraft that Jumbo Ashfield, now a Flying Officer with an AFC and a DFC, repeated his success of the previous July. He shot down another enemy raider on the night of 7 November 1940 and thus became the first pilot to shoot down two enemy aircraft at night. The Beaufighters slowly went into squadron use and the first kill by 604 Squadron followed a few weeks later.

As 1940 drew to a close, two factors brought about a complete change in the night interception scene. The first was that the Beaufighter gradually replaced the Blenheim in squadron service. The other was the introduction of the GCI radar into the ground control system. This was produced with commendable speed by AMRE, first at Dundee, then at Swanage, by a small team under the leadership of Denis Taylor, by modifing a CHL set to perform the ground control function. Fighter Command made an urgent demand for six of these sets to be completed by the end of the year. The first model went into operation at Durrington, near Worthing, in October 1940 and the very first controlled interceptions were made by the scientists who designed the equipment. Gradually, Air Force controllers took over and the remaining stations were completed by the end of the year. For the first time, the night fighters were being placed accurately on the tail of raiding aircraft and, when they got there, the Beaufighters had ample speed to complete an interception.

The transformation has been described by Group Captain John Cunningham, DSO and two Bars, DFC, one of Britain's most successful night fighters. In a quotation taken from Watson Watt's book *Three Steps to Victory*, he has described the change which came about in the following terms:

> . . . the Beaufighter had ample performance and adequate armament to deal with many of the German bombers then being used. . . . Not until the end of November did our hopes begin to rise that at long last we had got a combination of airborne radar and ground control that was adequate to handle an interception. The November success was very rapidly followed in December by a number of more successes. . . . It caught on like a snowball. . . . culminating in the spring of 1941 with very considerable successes in the night fighter squadrons.

In the first two months of 1941, the weather was unusually bad over Europe and this restricted enemy activity. In March, the enemy resumed heavy night raids on Britain and during that single month the night fighters scored 22 kills. In April, the number of enemy aircraft destroyed was 52, with a further 88 probably destroyed. In the month of May 1941, a total of 102 confirmed kills were made by the night fighters over British soil. In addition, no less than 172 enemy aircraft were severely damaged, many of which could never have reached their home bases. These were losses which the German Air Force could not sustain and from June onwards, apart from occasional flurries, there were few night bombing attacks of any significance on Britain. Once again, this success belongs to the aircrew and the many men and women on the ground who made it possible. An absorbing account of the final success can be found in the book *Cover of Darkness* by Group Captain Chisholm and in that other excellent work, *Night Fighter,* by Jim Rawnsley and Robert Wright, both of which have already been mentioned.

The German Air Force suffered its first great defeat in 1940, during the daylight Battle of Britain. In the spring of 1941, they were defeated again in the night battle, never to return in force against the British mainland. We will never know whether this was the event which persuaded Hitler to turn his attack in an easterly direction. However, one thing was clear. Any further attacks he cared to mount against the west would be met by a devastating response.

Centimetre-Wave Air-Interception Radar

As in the case of sea-search radar, the next step in the evolution of air-interception radar was the design of a set which operated on the extremely short wavelength of 10 centimetres. As we shall describe in later chapters, this arose from Boot and Randall's invention of the resonant magnetron. This was a device which made it possible to build an entirely different and vastly improved breed of air-interception equipment operating on centimetre waves. These new designs came from two sources, the Centimetre Wave Group at the Telecommunications Research Establishment (TRE) in Swanage and the Radiation Laboratory at the Massachusetts Institute of Technology (MIT) in the USA.

The development of AI radar at Swanage was the responsibility of Alan Hodgkin, who had joined the orignal airborne radar team at St Athan in 1939. He made his first flight with a 10 centimetre radar out of Christchurch on 10 March 1941 in one of the old Blenheims of 'D' flight. They obtained a range of two or three miles on a target aircraft and this was rapidly improved upon. This line of development led first to AI Mark VII and then to Mark VIII, both of which were installed in Beaufighters in the summer of 1942. Similar installations followed in the Mosquito and these remained in service until the end of the war.

An experimental 10 centimetre air-interception set was also built at the Radiation Laboratory of MIT and, by the summer of 1943, a production version manufactured by the Western Electric Company began to appear in Britain in the form of the SCR 720. This became known as the AI Mark X in the RAF. The main difference between the Mark VIII and the Mark X was in the scanning and display system, but the maximum range performance was comparable at six to eight miles. For the first time, this range could be achieved at any height above ground and was a vast improvement over the old 200 Mc/s equipment. Due to its superior performance against jamming from enemy sources, the Mark X gradually took over from the earlier versions of AI radar and, as production from the USA built up, it became the principal AI set in the Royal Air Force. It remained the standard AI equipment in the Royal Air Force until 1957.

Figure 7.1 A Bristol Beaufighter showing AI Mark IV antennae. (Courtesy of the Imperial War Museum.)

Figure 7.2 A de Havilland Mosquito fitted with AI Mark VIII. (Courtesy of the Imperial War Museum.)

The Mosquito in Intruder Operations

We have seen how the Beaufighter, on its first appearance as a night fighter, put on a spectacular performance and virtually won the night battle in the spring of 1941. The de Havilland Mosquito was an even better aircraft and, from 1942 onwards, gradually took over the role of night fighter. The Beaufighter moved on to other fields. It played a decisive role, both as a night fighter and a ground attack aircraft, in the later stages of the Mediterranean war. In the Pacific and in Burma, it became known to the Japanese as 'The Whispering Death' and excelled itself in low-level attacks against surface and marine transport.

As night raids on Britain tapered off, the radar-equipped Mosquitoes began to play an increasing role in assisting the Allied bombing attack on occupied Europe. They did this in many ways; they accompanied the bomber streams and protected them from attacks by German night fighters, which by that time had become a very formidable force. Perhaps the most important role of the Mosquitoes was as intruders. On the night of a big raid, they would hover around enemy airfields and attack the German night fighters when they were most vulnerable—that is, as they took off from their home bases. Another favourite time for attack was when the German night fighters were returning to base after a long raid. They might have already suffered damage or were tired from a long night in the air. They would certainly be relaxed by the fact that they had finally reached home. There are several cases of Mosquito pilots shooting down two enemy night fighters over a German base early in the night, returning to Britain to refuel and then going out to the same base and shooting down two more. In operations of this kind the Mosquito is credited with destroying over 600 enemy aircraft over German soil, and there were several squadrons with remarkable tallies of 50 or more kills per squadron. An engrossing account of these operations is given in a book by Squadron Leader Lewis Brandon entitled *Night Flyer* (1961). He was not a pilot but an AI radar operator and was awarded a DSO and a DFC and Bar and was among the RAF's most decorated radar operators.

As usual, the Germans were quite explicit about the havoc wrought by the Mosquitoes. Wilhelm Johnen, who finished the war as Commanding Officer of a German night fighter wing, said in his book *Duel Under the Stars:*

> . . . Mosquitoes from the mainland were despatched to join the bomber stream and take over the task of air cover. . . . They were the greatest plague and wreaked havoc among the German crews. The

> radar equipment of this wooden aircraft surpassed anything that had previously been seen. It was technically so perfect that at a distance of five miles they could pick the German night fighters out of the bomber stream like currants out of a cake . . . . The losses rose appallingly . . . The Mosquitoes (also) waited for us as we took off from our airfields. They attacked us throughout the whole operation and interfered with our landing.

The Mosquito against the V-1

Still later in the war, a new threat to British cities arose in the form of the V-1 flying bombs; but this form of attack was anticipated and several forms of defence were mounted against it. The V-1s released in daylight were attacked by Tempests which hovered over the launch sites along the French coast and attacked the V-1s as soon as they became airborne. At night, this role was taken over by radar-equipped Mosquitoes which waited for the tell-tale burst of flame from the ground which signalled the launch of another flying bomb. The Mosquitoes would dive in pursuit, using their radars to keep track and bring them within firing range. Some aspects of the operation were easy—for example, the V-1 flew a perfectly steady course and did not take evasive action. Secondly, it was easy to get a visual—even in bad weather, the exhaust flame could be seen from half a mile away. However, there was very considerable danger. The speed of the V-1 exceeded that of the Mosquito, so the only way to close on the V-1 was to have plenty of height and make a diving attack. Some of the Mosquitoes were specially boosted with ethanol to give them extra speed, but this was not popular with the aircrews. The Mosquito was amply strong for its normal role, but after such boosting the stresses were so great that the aircraft was in danger of disintegrating. There was another problem. When fired on, the V-1 had a nasty habit of blowing up with the full force of its 2000 pounds of high explosive. Several Mosquitoes were damaged in this way and some were lost. They quickly learned to avoid the dead astern position and make their attack from the beam—not the easiest proposition given the extraordinary speed at which they were flying.

The Tempests and Mosquitoes operated mainly over the launch sites, but those flying bombs which escaped their attention and crossed the Channel had another hazard to overcome. As we shall describe, a screen of SCR 584 gun-laying radars was deployed along the English coast and these wrought a heavy toll on the flying bombs which succeeded in crossing the Channel. It would be hard to distinguish between aircraft and the guns as the greatest destroyers of V-1s, but it is estimated that between them they cut down more than 50 per cent of those launched.

8

The Move from St Athan to Swanage

I shall always remember St Athan as the place where we performed an enormously complicated juggling act—one which was far more complex then anything ever seen in a three-ring circus. It was a situation in which, from sheer necessity and not always successfully, we tried to keep far too many balls in the air at once. This was done with totally inadequate resources and virtually no administrative back-up. Very little of the story exists in the official record and this may be a good moment to gather the threads together.

The main task at St Athan was handing over to the Air Force the task of fitting AI and ASV radars to their aircraft. From any point of view, this was enormously successful. Within a few months of our arrival, St Athan was fitting one aircraft a day with either air-interception or sea-search radar and, as time went on, these figures were far exceeded. Even more striking than this was the AI and ASV production schedule from E K Cole and Pye Radio. Towards the end of 1939 and into 1940, Touch was spending most of his time conferring with the manufacturers. In February 1940, he moved his headquarters to RAE Farnborough, where the facilities for liaison with contractors and aircraft manufacturers were very much better than those at St Athan. Orders were placed for the production of 3000 each of AI and ASV sets, and by the end of 1940 the flow was under way. Production of radar equipment at a rate of nearly 10 a day, and a fitting programme to match, represented an entirely new dimension in the production and deployment of radar equipment; it did not occur to us that this was anything out of the ordinary.

Touch's first project was a properly engineered version of ASV radar—this was the model which became known as ASV Mark II and the one which played such an important part in turning back the submarine attack in the Atlantic. It became one of the most versatile radars ever built; it was produced in quantity in Great Britain, the USA, Canada and Australia and was used in every theatre of war. It probably saw more operational use than any other single piece of radar equipment. We have already mentioned how it spawned the Coast Defence, or CD, radar, which in turn led to the Home Chain Low (CHL, a network added to give warning of low-flying aircraft) and then to the GCI radars, all operating within the 200 Mc/s frequency band. Still later, ASV Mark II was used by the Army as the basis for a searchlight-director radar—the SLC. By an amusing twist of fate, it was also copied by the Japanese from a working model of the SLC which they collected at Singapore. Parts of ASV Mark II were also used as the airborne component of the OBOE blind-bombing system—a device which was used with enormous success in the bombing of Germany by the Pathfinder Force of the RAF. The engineered version of ASV Mark II was soon followed by the AI Mark IV. This was the set which had the greatest number of successful night interceptions to its credit over British soil. (See Appendix 1 for production figures of both AI and ASV equipment.) If for nothing else, the St Athan period will be remembered for the start of mass production of airborne radar, and for the fitting of these sets to aircraft at an unprecedented rate.

Yet another outcome of that period was the beginning of centimetre-wave work in Britain. The whole of this development took place at other establishments, but the stimulus came from the urgent need for an airborne radar with a good maximum range performance against low-flying aircraft. This requirement was recognised well before the outbreak of war, but as we shall describe in the next chapter, the steps towards solving it through centimetre waves did not gain real momentum until the spring of 1940.

As the work of the airborne group at St Athan drew to a close, yet another move was contemplated. From every point of view, the sensible thing for us to do was to go back to Bawdsey and Martlesham where many of us still had roots and where the facilities were vastly superior to anything we had experienced during the first few months of the war; but the bureaucrats did not see it that way. What happened was really decided by events at Dundee, where things had reached crisis point. The fiction that Dundee was a good place for radar research could no longer be maintained and another move simply had to be made. In describing these events in his book *One Story of Radar*, Rowe said:

> . . . it is difficult to find good in the Dundee era . . .If however we were seemingly bankrupt it is not to be supposed that grand work was not done elsewhere.

He was presumably referring to the field work done away from Dundee, even the work at St Athan. He went on to say that it:

> . . . was the last place to choose for radar research and all concerned with selecting it or not opposing the Dundee site, including myself, had learned a lesson not to be forgotten.

This was a generous admission, but it does not alter the fact that the move to Dundee was a disastrous mistake.

The Germans were now streaming through the Low Country, while in Britain things were in a chaotic state. Possibly for good reasons, the Air Ministry was split into two parts, the Air Ministry and a Ministry of Aircraft Production, and it was some time before equilibrium between the two groups was established. There were new bosses everywhere, many with little or no knowledge about what had happened in the previous few months, let alone the previous few years. In despair, Tizard wrote to the Chief of Air Staff and all three Commanders-in-Chief of Fighter, Bomber and Coastal Commands, saying that there were now so may advisers available—all giving contrary advice—that he wished to resign as the official scientific adviser. This led to such unanimous expressions of support for him from the Air Force and expressions of outrage from the scientific community that, reluctantly, he was persuaded to continue.

Meanwhile, a new site for AMRE was selected near Swanage on the south coast. I have never heard the reasons why that choice was made but, once again, it turned out to be a singularly inappropriate place. The prime reason for the disorderly flight from Bawdsey to Dundee was an obsessive desire to get away from the enemy. If that was still the criterion, the architects of the move from Dundee were wrong for a second time. By the time AMRE had settled at Swanage, the Germans had moved even faster and were just 60 miles away on the Cherbourg Peninsula. The site at Swanage consited of a series of unfurnished wooden huts with no facilities, without heating and surrounded by a sea of mud. P I Dee, an elite member of the Cavendish laboratory, had now joined the staff and his diary gives a graphic account of his first day at Swanage on 15 May 1940.

> . . . First day at AMRE . . . No-one seems particularly interested in our arrival. We were directed to a hut in the middle of a field where there was one desk, one stool and a filing cabinet. We take it in turns to sit on the stool, or alternatively, on the coke stove which is never lit. There are no facilities whatever.

At the same time, the civil aerodrome at Christchurch became the new home for 'D' flight and this was an equally inept choice. To anyone with the least experience of flying, it was totally unsuitable for the work in hand, but as was customary in those days, we were not even consulted. Christchurch had a single grass runway, too short for safety. It was surrounded on all sides by domestic residences which prevented any possibility of extending the runway. At the western end was a tennis court which was quickly demolished by a Hurricane which failed to clear the wire surround. If nothing else, that meant one less Hurricane to contribute to the coming Battle of Britain. The only building available at Christchurch to shelter both the air and ground crews for about 20 aircraft consisted of a 20 × 20 foot wooden hut. There was no place at all for the testing of new radar equipment going into aircraft. Compared with Bawdsey and Martlesham, it was a ludicrous choice. We had already experienced totally inadequate facilities at Perth and St Athan. Why this was repeated for a third time at Christchurch passes understanding. The simple fact was that we were now in the hands of people who had no conception whatsoever about the requirements for sophisticated radar research. Britain's war effort suffered enormously and it was a long time before this situation was rectified.

From sheer strength of personality, Dee rapidly became a powerful force at Swanage, now renamed the Telecommunications Research Establishment. Among other things, he made himself acquainted with the recent history of airborne radar. His diary goes on to report:

> . . . Bowen has had severe political rows with Rowe—I think because Bowen exposed the lack of interest by Rowe and Lewis in the AI work which he and his group was doing at St Athan. Tizard became wise to the situation and visited Dundee a short while back when he, Tizard, beat up Rowe for his lack of interest in AI work.

As a member of the old band, to whom the halcyon days of Orford and Bawdsey were still a vivid memory, my views about Swanage were not any different from those of the new arrivals from the universities. I can remember a cold and miserable morning, soon after my own arrival at Swanage. There was a steady drizzle coming in from the west; in raincoats and gum boots, which were essential in the prevailing mud, Hanbury Brown, Hodgkin and myself were sitting on a soaking wet plank supported by a few bricks under the eaves of a tumble-down fowlshed in the bottom corner of the new establishment. A local farmer's wife had taken pity on the staff and was determined to do her bit for the war effort. She had refurbished the hut to dispense tea, at a penny a cup, to the freezing and miserable inhabitants of the new Air Ministry Establishment.

We sat there, with rain dripping down our necks, clutching cups of tea in both hands to provide a little warmth, thinking how different this was from the happy days at Bawdsey. After a few moments Hanbury said: 'Holy Moses! If Rowe could only see us now. Do you know what he would do? He would write a Station Order saying we should not be doing this.' Hanbury could mimic Rowe's style to perfection, and he proceeded to compose the Station Order right there. Within two or three days, it appeared in almost exactly the words used by Hanbury. an exact copy of Rowe's memorandum follows:

Dr. Bowen

USE OF CANTEEN

I was astonished yesterday to find that a considerable number of the staff are using the Canteen for tea and other refreshments in the middle of the morning. No man in normal health needs refreshment in a morning so short as one from 8.45 a.m to 1 p.m., and in general the Canteen is not to be used in the morning. This instruction needs, of course, to be interpreted with common sense. If a member of the staff is feeling unfit or has been working out in the wet or cold, there is no objection whatever to the use of the Canteen.

2. I think everyone knows that I am strongly in favour of tea in the afternoons. In most cases, however, it will save time if the tea is made in the Laboratories and other buildings by the junior member of the staff concerned, or by one of the very willing members of the typing pool.

3. Please circulate to every member of your staff.

13th July 1940

Superintendent.

The staff had met this kind of eccentricity before and, in their strictly practical way, they solved it by switching to coffee in the morning and sticking to tea in the afternoon.

Due to the unsuitability of Swanage, the establishment had to survive yet another move, to Malvern in the Severn valley, before it finally settled down. By comparison with this bewildering series of moves, first from Bawdsey, then from Dundee, and finally from Swanage to Malvern, it is worth making a comparison with the Royal Aircraft Establishment at Farnborough, an Air Ministry Establishment strictly comparable with TRE. In 1940 it occupied a site well known to the enemy and easily recognised from the air just outside London; but it stayed in the one place for over 50 years and for the duration of two wars.

It has been said that the move to Swanage was the turning point in the fortunes of an establishment which, at Dundee, had run down almost to the point of extinction. After the final move to Malvern, there is no doubt that it became one of the great radar research establishments of the war. History alone will decide the factors which brought this about. It could be that after a sequence of disastrous moves, the management had finally learned how to avoid catastrophe. With the benefit of hindsight and 50 years' experience of radar research in several continents, I can still recall no good in the management; to me, the only reason TRE went on to greater things was the superlative dedication of the staff who managed to survive the appalling conditions under which they were expected to work.

At this tumultuous stage in the history of the development of British radar, I was extremely lucky to be selected by Sir Henry Tizard to continue my career in a somewhat different role.

9

The Transition from Metre to Centimetre Waves

We now know that before the outbreak of war several countries were ahead of Great Britain in the construction of centimetre-wave radars. For example, the German Navy demonstrated a continuous-wave ship-detection system at Kiel in 1934, working on a wavelength of 13.5 centimetres. In 1935, the passenger liner *Normandie* carried a very similar 16 centimetre system which indicated that there were icebergs ahead. This latter was no secret and was fully described in many newspapers at that time.

To be perfectly fair, these were not true radio location systems but obstacle detectors. They showed that there was an object somewhere in the vicinity, but they gave neither its position nor its distance. But proper position-finding was not far behind and a pulsed version of the Kiel equipment was shown to Admiral Raeder in May 1935. This demonstration of ship detection took place just one month before the demonstration of pulsed aircraft echoes at Orfordness in June 1935. The German system had a maximum range of detection of ships of 8 kilometres, but to get the requisite performance the wavelength had been pushed up to 48 centimetres. In 1938, Dr Ponte, in collaboration with the French Navy, built a pulsed version of the *Normandie* equipment at Le Havre. It operated on the original wavelength of 16 centimetres and gave a range of 3 kilometres on fishing vessels and 8 kilometres on larger vessels. By comparison with this, in 1938 the range of detection on a capital ship with British 200 Mc/s sideways-looking radar was already close to 20 miles and it is possible that the very success of this equipment may have delayed a move to shorter waves in Britain.

The trend to centimetre waves came very slowly in Britain, but when the change came, it did so with dramatic speed. A single event led to this revolution—the invention of the resonant magnetron by Boot and Randall at Birmingham University in 1940. Britain shot into the lead in centimetre-wave radar and, once in the lead, she maintained it quite easily until the end of the war.

To understand how the change came about, it is necessary to go back a little in time to the very beginning of radar in Britain. Since the first success at detecting aircraft with radio pulses at Orfordness in 1935, the trend in operating wavelength was steadily downward. Starting at 50 metres, the wavelength was changed first to 26, and then to the band from 10 to 13 metres. Experimental systems were also built at Orfordness and Bawdsey at wavelengths down to 4 metres, followed in the summer of 1937 by widespread use of 1.5 metres for airborne and other applications. With the components available in Britain at that time, both the transmitted power and the receiver sensitivity of a radar system fell away in an alarming fashion at wavelengths below 1.5 metres and for a while this became the lower limit. However, we all dreamed of shorter wavelengths and narrow beams; a copy of the first page of a memorandum I sent to Watson Watt dated 9 October 1935 is given in Appendix 2. It mentions an attempt to develop a 50 centimetre pulse transmitter at Orfordness which, in the event, was not successful.

The memorandum then goes on to enquire about the possibility of acquiring 18 centimetre equipment from the Standard Telephone Company. This was a reference to an 18 centimetre communication link across the English Channel, built a few years earlier by that company. It used Barkhausen–Kurz oscillators of rather low efficiency and parabolic dishes about 20 feet in diameter. It was moderately successful as a communication system, but was not as reliable as a submarine cable, so the scheme was abandoned after a few years. Through Watson Watt, I made strenuous efforts to acquire this equipment to see if it could be pulsed and used as a detection system; but even the persuasive powers of Watson Watt could not extract the equipment from its owners and the matter dropped. If we could have obtained the 18 centimetre equipment at that time, the whole history of ship and aircraft detection in Britain might have been different.

In September 1938, my airborne radar group at Bawdsey constructed a complete 30 centimetre pulse detection system using Samuel 2B250 valves from Western Electric in both the transmitter and the receiver. As we were thinking only of the airborne application, the antennae were quite small and consisted of cylindrical parabolas two foot square. With these, the range of detection on ground objects turned out to be only three-quarters of a mile. This was small compared with our existing 1.5 metre systems and the project was not continued.

In the next few years, there is no doubt that it was the Admiralty which expressed greatest enthusiasm for changing to centimetre waves. There was practically no interest within Bawdsey itself and anyone talking about centimetre waves was thought of as some kind of crank. In the spring and summer of 1939, I discussed the problem a number of times with Sir Charles Wright, Director of Scientific Research for the Admiralty, during his many visits to Bawdsey. In addition to his Naval responsibilities, he was Chairman of the Committee for the Coordination of Valve Development (CVD), which acted on behalf of the Army, the Navy and the Air Force. We agreed that the greatest need for a centimetre-wave system was in airborne radar and he asked what wavelength was most desirable for this application. I did a back-of-an-envelope calculation of the best wavelength for the purpose. As we have seen, by far the most limiting factor in airborne radar at 200 Mc/s was the enormous signal which came from the ground directly below the aircraft. Aircraft echoes beyond that range would simply be wiped out by the much stronger signals coming from the ground. The only way of improving the system would be to project a narrow beam forward to get rid of the ground returns. I estimated that a beam width of 10 degrees was required to achieve this. Given an aperture of 30 inches—the maximum available in the nose of a fighter—this called for an operating wavelength of 10 centimetres. This agreed with Sir Charles's own assessment of what was required for Naval purposes.

It was not long before the CVD Committee placed contracts with Professor Oliphant at Birmingham, with Bleaney at Oxford and with Bristol University for the development of transmitting and receiving tubes for a 10 centimetre wavelength, and these became the centres at which centimetre-wave research blossomed. The name of Oliphant's laboratory, previously known as the Mond Laboratory, was changed to the Admiralty Research Laboratory, Birmingham University. This happened towards the end of 1939 and, as we shall see, it was the crucial step which led to the invention of the resonant magnetron. A noteworthy fact is that the staff of Oliphant's laboratory were mostly nuclear physicists, not electronic engineers. Oliphant correctly claimed that this was a great advantage. By virtue of their training, the staff would not be inhibited by prior convictions about what could and could not be done.

About the same time, interest at the Air Ministry had grown to a point at which a development contract was placed with the GEC Research Laboratory at Wembley for an airborne radar system on centimetre waves. The exact wavelength was not specified, but it was intended that it should be around 20 or 30 centimetres. This was based on the fact that GEC had made a significant improvement in the performance of conventional triodes and already had a valve—the VT90 micropup—which could generate many kilowatts at a wavelength of 50 centimetres. They had hopes that, with further development, a valve which they called the millimicropup might be squeezed down to a wavelength of 20 centimetres. This was getting close to the point where it would be extremely useful in many airborne applications.

The Air Ministry was keen to explore this possibility; as always, it was Tizard who took the lead and he visited the GEC Research Laboratories at Wembley on 6 November 1939 to take the matter further. He stressed the importance of short-wave work and pointed out that, due to lack of facilities for valve making in most government establishments, it was essential that work of this kind be undertaken by industry. This was followed by a visit from Watson Watt when he discussed a form of contract with GEC. As a result, the Director of Communictions Development (DCD) formally asked GEC to accept a development contract for a short-wave air-interception radar on 29 December 1939. The EMI Company at Hayes was asked to collaborate in the same contract and a series of joint meetings took place commencing on 16 January 1940. I was asked to act as coordinator. Meetings were held first at monthly intervals and then fortnightly. They were held alternately at the GEC Research Laboratories, Wembley, and at EMI, Hayes. Dr C C Paterson acted as Chairman at Wembley and Dr Schoenberg at Hayes. Prominent members of the committee were Gossling, Megaw and Jesty for GEC and Blumlein and E L C White for EMI. The purpose of these meetings was to arrive at the design of a prototype airborne radar on centimetre waves with which air trials could be carried out.

In many ways, the placing of this contract was a unique event. It was probably the first occasion on which a developmental contract—as distinct from a production contract—was placed for radar equipment in Britain. Up to that time, the number of companies involved in radar production had been severely restricted because of security considerations. The main purpose of this contract was to stimulate work on centimetre waves, but it was also a deliberate effort by the Air Ministry to widen the number of companies involved in radar production. We were particularly impressed with Megaw's expertise in valve design—we could clearly have drawn on his talents much earlier. Blumlein soon proved that he was a master of circuit design and produced a flow of innovative ideas which transformed circuit design during the next few years.

A record of the minutes of the above meetings was kept by the GEC secretariat and probably still exists in the official record. For those interested in the details, a list of the dates of these meetings from my diary is given in Appendix 3. At the first two meetings, I described the 200 Mc/s airborne systems then going into operational use and how these could be improved by the use of shorter wavelengths. Central to the discussion was the belief that maximum benefit would not be obtained unless a wavelength of 25 centimetres or less was achieved. A wavelength of 50 centimetres, while extremely useful for many other applications, was undoubtedly too long for airborne use. It would simply not provide a beam width narrow enough to get rid of the ground return. By operating at that wavelength, one would end up with a system which had a poor performance compared with existing 200 Mc/s equipment.

Much of the discussion revolved around antennae—parabolas versus horns. Based on prior American experience, horns were favoured at first, but the GEC engineers soon produced evidence that paraboloids were more efficient. There was also an active discussion of scanning and display systems for air-interception and sea-search radar respectively. Another great attraction of a centimetre-wave system was that it would lead to a dramatic improvement in the way information was displayed to the operator. For the first time, the display could be in pictorial form, the most spectacular of which was the Plan Position Indicator, or PPI, which was already under development.

As always, the ship-detection application was seen to be the easy one—scanning would be by simple rotation of the antenna about a vertical axis combined with a PPI display system which gave range and bearing pictorially. We thought in terms of a six foot cheese-shaped antenna rotating continuously under the belly of a reconnaissance aircraft. The AI application required scanning in elevation as well as in azimuth and was, therefore, more difficult. The most favoured system at that time was a helical scan system which I put forward, with 360 degree rotation in azimuth combined with a limited scan in elevation. A sketch of the proposed display system taken from a contemporary document is given in figure 9.1. This shows three two-dimensional displays. These are the range–azimuth display, or so-called B scan, the elevation–azimuth display, which became known as the C scan, and the range–elevation display, or E scan. Any two of them would supply the information needed for an airborne interception; the B and C scans were the ones ultimately used in AI Mark X. This shows the way we were thinking early in 1940 and it remains the basic method of display used in most airborne systems to this day.

In May and June 1940, a complete radar was operating on the roof of the Wembley Research Laboratories at a wavelength of 25 centimetres. It gave echoes from ground objects at a range of 5 or 6 miles, which was already competitive with similar systems working on

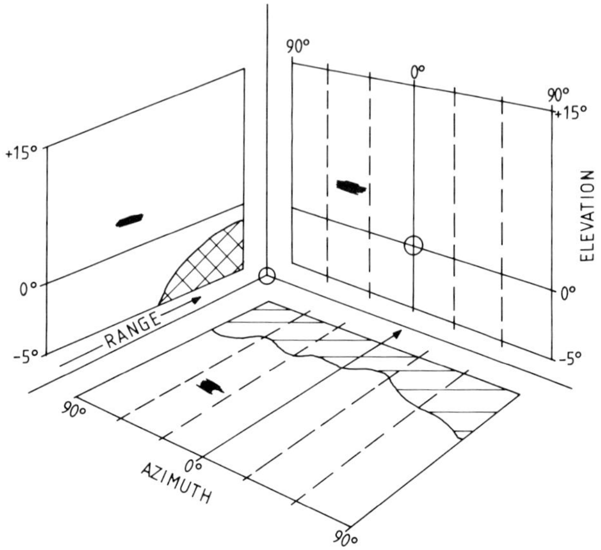

Figure 9.1 The method of displaying range, azimuth and elevation in a centimetre-wave airborne system. Ground echoes are shown shaded and the dark area is an aircraft echo at 45° to port and +5° elevation. Redrawn from a 1940 document

200 Mc/s. With this kind of progress being made at Wembley, some notice was at last being taken at Swanage. On 8, 10 and 12 June 1940, I gave a series of lectures to some of the new recruits on these developments. Even so, not everyone was convinced that the centimetre-wave route was the way to go. In his book *One Story of Radar* Rowe said of the period:'. . . at least one voice warned me almost daily of the folly of rushing into a centimetre application'. But the centimetric revolution was well under way and the management of TRE could ignore it no longer.

It was at this point that news of Boot and Randall's resonant magnetron came through. Back at Birmingham, it was initially thought that the best hope of developing power at the centimetre-wave range lay in the klystron, invented a few years previously by the Varian brothers at Stanford University. Oliphant had seen an early version of this device during a visit to the United States in 1938. It was already in use as a low-power oscillator in the USA, but no high-power designs had been attempted. A major effort went into producing a high-power version of the klystron at Birmingham and some progress was made. Randall, who had been on another assignment, was late joining the group. When he did so, he joined forces with Boot and was asked to start on the much less promising task of trying to increase the output of the magnetron, which up to that time was known to be a very inefficient device.

Randall has told a fascinating story of how he hit on the way to improve the magnetron. He went back to the beginnings of radio and read up Hertz's original experiments. Hertz's receiver consisted of a single resonant ring with a small gap across which a spark jumped when the transmitter was keyed. In his mind's eye, Randall imagined the Hertz oscillator extended into a cylinder with a slit in it, as shown in figure 9.2 (*b*). He then considered how this might be used in a magnetron and, in a flash of inspiration, saw that the way to go was to use a number of resonators arranged around a circle as in figure 9.2 (*c*). This drawing is derived from Randall's actual sketches made a few months before he died in 1984.

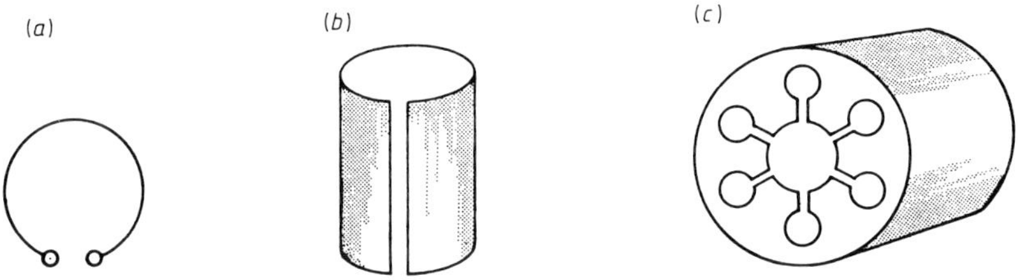

Figure 9.2 How the Boot and Randall resonant magnetron evolved. (*a*) Hertz's original resonant ring. (*b*) Its extension into a resonant cylinder. (*c*) How this developed into a six-cavity system in Randall's mind. This drawing is taken from sketches made by Randall just before his death in 1984.

So the principle of the resonant magnetron was born and Boot and Randall immediately set about constructing a working model. Legend has it that the first core was drilled out of a copper block using the barrel of a Colt revolver as the jig, but there is uncertainty about this. The first model was far from being the finished version. It was strictly a laboratory model, requiring a tangle of glass tubing and pumps to maintain a vacuum, but it produced power for the first time on 21 February 1940. It ran first as a continuous-wave oscillator and immediately showed great promise. Shortly afterwards it was pulsed, with dramatic results. The pulsed output was well over one kilowatt at a wavelength of 10 centimetres and it was apparent that a gigantic breakthrough had been made.

Contact was then made with Megaw at GEC and discussion started on the design of a compact, sealed-off version of the magnetron which could be used in an aircraft. Megaw made two vital contributions. Firstly, he suggested the use of an oxide-coated cathode, something which he had just heard about from Henry Gutton in France. Secondly, he suggested the gold seal, which allowed compact end-plates to be attached to the magnetron and the final seal made. Incorporating these two improvements, the first production models began to come through the shop at Wembley in July 1940 and their performance was outstanding. With a magnetic field of 1500 gauss and a pulsed anode potential of 10000 volts, they gave an output of 10 kilowatts, with a distinct possibility of more to come. After 45 years it is difficult to reproduce the drama of the occasion, but the performance of the resonant magnetron was simply revolutionary. The power it produced was already greater than that obtained from conventional airborne transmitters on 1.5 metres and it would clearly lead to a complete revolution in every kind of radar, ground-based, shipboard and airborne.

During the last week of July, the GEC roof set was converted to operate either on 25 centimetres with micropups, or on 10 centimetres with the resonant magnetron. Comparative tests were carried out and it was obvious that the 10 centimetre wavelength out-performed the 25 centimetre wavelength by a wide margin. This was the most important item of discussion at the regular meeting with GEC on 5 August 1940 and, from that moment onwards, the project settled on 10 centimetres as the working wavelength. A Centimetre Wave Group was then formed at Swanage under the dynamic leadership of P I Dee. It included people like Lovell and Hodgkin from St Athan, who were already familiar with airborne radar at longer wavelengths and therefore had a head start in the design of the components for such a system. Like Oliphant, this group had also focused its attention on the klystron in the first instance, and they had one working on the pumps at Swanage. When the sealed-off resonant magnetron appeared, this was clearly the way to go. During the early part of August, the Swanage group assembled a laboratory system using a GEC magnetron and reproduced the Wembley results on ground echoes. Towards the end of the month, they obtained the first aircraft echoes from one of our Battle aircraft at a range of one or two miles. From that moment, the success of 10 centimetre radar was assured. Following the formation of the Centimetre Wave Group at TRE, similar groups were set up at the Admiralty and at the Army Air Defence Research and Development Establishment (ADRDE) to undertake development of shipboard and gun-laying radars for Naval and Army use respectively.

Meanwhile, another profound change was in the wind. Tizard had been active in yet another direction. As we shall describe in the next few chapters, he made arrangements for British top secret developments, including the magnetron, to be transferred to America to take advantage of the research and production resources of that country. I was soon fully involved in that effort. From that point onwards, the development of centimetre-wave radar by the Allies followed two distinct, but parallel, paths—one in Britain at TRE, ADRDE and the Admiralty, and the other in the United States.

Work at GEC and EMI went on and the joint meetings continued, with Tommy Walmsley of the Ministry of Aircraft Production (MAP) as Chairman and Dee as the representative from Swanage. The Valve Laboratory at Wembley undertook further development of magnetrons of ever increasing power and operating over a wide range of wavelengths, while EMI, with Blumlein in the lead, specialised in advanced circuit development. The Centimetre Wave groups at TRE, ADRDE and the Admiralty gradually took over responsibility for the design of complete radar systems while the GEC and EMI companies took on production and engineering responsibilities. These two companies played a profound part in the initial development of centimetre-wave radar in Britain, and they have continued in the forefront of research and development to this day.

10

The Tizard Mission to the USA and Canada

By the early months of 1940, Britain was in desperate straits. The big attack on Britain and France was yet to come, but nobody was hopeful about the outcome. Tizard foresaw that it would probably not be long before the United Kingdom would be cut off from the Continent and without allies. Left to her own resources, she would soon be near the limit of her productive capacity, particularly in the all important field of electronics. Even if the United States did not enter the war, it would become essential for Britain to invoke both the technical resources and the productive capacity of the North American continent. This process had already started with conventional weapons like guns, tanks and aircraft, but Tizard was thinking of more exotic developments, particularly those in air defence where Britain might have a head start over the USA.

Quite early in 1940, therefore, Tizard made the bold suggestion that Britain should disclose her secrets to the USA in return for help on technical and production matters. At first, this proposal did not meet with a very enthusiastic response. Watson Watt, for example, is on record as saying that both the research and the productive capacity of the United Kingdom were adequate and that the USA 'had nothing to offer'. Sir Frank Smith, who was then the Comptroller of Telecommunications Equipment, expressed himself as being 'constitutionally opposed' to invoking aid from the USA and pointed out the danger of loss of secrecy if this course were adopted. Even Churchill, then First Lord of the Admiralty and a rising power in the land, was not entirely convinced and did not think the time was ripe.

As usual, Tizard was not afraid of advancing a point of view which might prove to be unpopular; in fact, the suggestion that Britain's research and production resources might not be adequate was resented by some of his colleagues. He was not discouraged by this and went quietly to work. As a preliminary step he arranged for his colleague A V Hill to go to the British Embassy in Washington, with the task of enquiring as discreetly as possible into the state of military research in the USA, especially in air defence. Professor Hill did not have the authority to disclose British advances in this field and, as a result, he did not learn much about American activities. However, he was convinced that there was an immense fund of goodwill towards the British cause in American scientific circles and that a great potential for research, development and production existed in the USA and Canada, provided the right approach were made.

He sent this encouraging news back to Tizard; but, in addition, he made another most important recommendation—that no attempt should be made to trade secret for secret with the Americans. Subject to appropriate security arrangements, he advocated that there should be a complete disclosure of British defence secrets up to that date. He was confident that the Americans would react to this in a generous and wholehearted way. Tizard took to this suggestion at once and renewed his advocacy of the proposal.

By now, France had fallen. With German forces encamped along the Channel coast and with German submarines occupying bases along the Bay of Biscay, it was a good deal easier for Tizard to get his point across. Churchill had become Prime Minister and was now a strong supporter of Tizard's proposal. He spoke to Roosevelt directly about the need for a mission to the USA. Even Watson Watt was converted to the idea, so much so that he himself wanted to act as leader of the mission. Final negotiations for the visit took place between Lord Lothian, the British Ambassador to Washington, and Roosevelt in mid-July 1940. (See an *aide-mémoire* from Lord Lothian to Roosevelt dated 8 July 1940 in Appendix 4.) So fast did things move that the proposition was approved by both Governments by the end of July and the members of the mission were selected in the early part of August. They left for the United States on 29 August, arrived in Halifax, Nova Scotia, on 6 September and began operations in Washington a few day later.

The members of the Tizard Mission were:

Sir Henry Tizard (Mission Leader)
Brigadier F C Wallace (Army)
Captain H W Faulkner (Navy)
Group Captain F L Pearce (RAF)
Professor John Cockcroft (Army Research)
E G Bowen (Radar)
A E Woodward Nutt (Secretary)

The Service representatives were all senior people with recent operational experience. Brigadier Wallace, an MC from World War I, had been in charge of the anti-aircraft defences at Dunkirk and was one of the last men to leave the beaches. Captain Faulkner had served in the Atlantic and took part in the campaign in Norway. Group Captain Pearce had made the first of many bomb attacks on the *Scharnhorst* in a Norwegian fiord.

Cockcroft, before going on to a distinguished scientific career at the Cavendish Laboratory at Cambridge, had served as an artillery officer in World War I and had recently returned to the Army to advise on a wide range of Army research projects. On his return from America he was due to be appointed head of ADRDE, the Army radar research establishment at Christchurch. I was the radar man, with experience of both ground and airborne radar.

The purpose of the Mission, subject to carefully vetted security procedures, was to hand over to the US Services all the recent British technical advances. These included things like the jet engine, then in embryonic form, rockets, predictors and, most important of all, radar in its many forms. I do not know of anything that was excluded. Tizard made the issues clear to us during meetings in London when he assigned our individual tasks. He left us in no doubt about the importance of the Mission, and the seriousness with which it was regarded by the Government. He also showed his wry sense of humour. Without diminishing the magnitude of the venture in any way, he hinted that between negotiating sessions in the US there would be some equally serious socialising to do. Looking at us in his whimsical way, he indicated that it was the job of the military members of the Mission to shoulder that particular burden. Would the civilian members please exercise restraint and keep their heads clear!

The broad plan was that Tizard would go ahead by air to make preliminary arrangements, while the rest of the party followed by ship. During those first hectic weeks of August, we collected all the documentation we could—manuals, circuit diagrams, blueprints, films—anything we could lay our hands on which gave a factual account of what was being done and what was projected. These were deposited in a black deed box of the kind found in a solicitor's office, which was kept under close guard in the headquarters of the Department of Supply at Savoy House in London, where Cockcroft had an office.

Easily the most important item in the box was a sample of one of the first production magnetrons from GEC at Wembley. This was the resonant magnetron just invented by Boot and Randall at Birmingham University; as we have described, the first twelve production models had been made by Megaw at the GEC Research

Laboratory only a few weeks before. I kept closely in touch with developments at Wembley and went there again on 7 August to have a detailed briefing on how the magnetron was constructed. Not only did it promise to revolutionise my own field of airborne radar, but it had an application for every variety of radar. A sample magnetron had already been dispatched to TRE in Swanage and another to the Signal School at Portsmouth. Megaw ran the remainder through the test rig and kindly let me pick the best of them to take to the USA. It happened to be Number 12, which was to have an interesting consequence. I returned to Wembley on 11 August to pick it up, together with a detailed account of the method of construction and blueprints giving the internal dimensions. It was customary for us to travel out to Wembley by underground, and that was the way I returned with the magnetron—no security guard, no staff car—to deposit it in the black box at Savoy Hill.

I spent the last days of August at the Air Ministry making final arrangements for the trip. I was to be the custodian of the black box on the journey from London to Liverpool so, late on the evening of 28 August, I went to Savoy House to pick it up, now fully loaded with Britain's most important secrets. It was delivered to me by the guard on the back steps of Savoy House. I called a taxi and took it to the Cumberland Hotel where I was staying. I had previously arranged with the hotel manager that he would keep the box overnight in the hotel safe. When he saw it, he raised his hands in dismay; it was bigger than his safe and there was no way he could store it overnight. There was no alternative; it had to go under the bed in my room where it remained until next morning, when I was due to catch the 8.30 boat train for Liverpool from Euston Station.

So, next morning, it was another taxi ride from the Cumberland Hotel to Euston. For reasons best known to himself, the taxi driver refused to let me carry the box inside the taxi and insisted that he should strap it on the roof. There was no time to argue, so we made the short run to Euston with that supremely important piece of luggage prominently displayed on the roof. With my other luggage, the box was more than I could handle, so I called a porter and told him to head for the Liverpool train. He grabbed the box, put it on his shoulder and headed off so fast that (an old cross-country runner and still pretty fit) I had great difficulty keeping up with him. He got well ahead and the only way of keeping track of him was to watch the box weaving its way through the mass of heads in front. A first class seat had been reserved for me, but beyond that I did not know what to expect. All the other members of the Mission were going to Liverpool by different routes, so I was alone on this leg of the journey.

I found my seat and, with the black box parked on the luggage rack, waited for departure time. What I had not realised was that the whole compartment had been reserved and when I entered it all the blinds were drawn and large notices were posted on both windows. A few minutes before departure time, an exceptionally trim, well dressed gentleman with a public-school tie came into the compartment and with scarcely a glance in my direction settled down in the seat diagonally opposite reading *The Times.* Not long after the train started, the late arrivals started shuffling up and down the corridor, looking for an empty compartment. A couple of bright sparks opened the corridor door and said 'Here we are chaps, this one is nearly empty' and started to enter. My companion spoke up for the first time and said 'Out. Don't you see this is specially reserved?' It was not what he said but how he said it. The would-be intruders wilted and we had no further interruptions. For the first time, I realised that the precious cargo was under some kind of protection.

In due course, we arrived at the dockside in Liverpool. My instructions were to stay in the compartment until the box was picked up by the Army. My companion also stayed put, absorbed in his *Times*. I became aware of the sound of marching feet and noticed that a squad of a dozen fully armed soldiers was marching down the platform, led by a sergeant. They came to a halt alongside the compartment and, with much slapping of rifle butts, went through a complicated set of manoeuvres which faced them in the opposite direction. They stood at ease. On a command from the sergeant, one of the squad broke away and opened the compartment door. Two more entered the compartment, collected the black box and trundled it outside. On a further word of command, they shouldered it and marched off in the direction of the ship. I was beginning to feel that things were well looked after. Alternatively, if this was the enemy making off with Britain's secrets, they were making a spectacular job of it.

My companion had not moved a muscle. When the squad finally departed, he rolled up his newspaper and with a barely perceptible nod in my direction, took his departure into the corridor. I had no doubt that he was heading for a well earned drink. Mission accomplished.

I was left to find my own way to the ship—the *Duchess of Richmond*—in which the remainder of the Mission was to cross the Atlantic. I arrived on board about 1.30 PM and verified that the black box had indeed been delivered and was under guard on the bridge. The Captain's instructions were that in the event of an enemy attack and a likelihood of the vessel being lost, the box was to be heaved over the side and allowed to sink in mid-Atlantic. We left our berth as darkness fell and headed for the Irish Sea. We had not gone very far before bombs started to fall around us. They were not particularly accurate, but were sufficient to stop the ship. There had been sporadic night bombing in London during the preceding few weeks, but no one took it seriously and this was presumably more of the same directed at Liverpool. We spent the night in the Roads and started off next morning. The *Duchess* was fast enough to make the crossing unescorted, making regular course changes every 20 or 30 minutes. She was called the 'Drunken Duchess', in view of her tendency to roll all over the high seas.

There were only two events during the passage which are worth recording, both involving Cockcroft. The only person with authority to have access to the black box was Woodward Nutt as Secretary of the Mission. At our request, he would occasionally extract documents for study. He often remarked on the difficulty of moving the box from its hiding place on the bridge. This led Cockcroft, in the bar one evening before dinner, to make the kind of calculation only Cockcroft would think about making. If, in the event of enemy attack, the box were thrown overboard, would it sink or would it swim? The answer was open and shut. The *Duchess of Richmond* would almost certainly go to the bottom, but the box and all it contained would float!

The second episode has its amusing side. Also on board the *Duchess* were well over a thousand officers and men of the Royal Navy, and it was not clear to us, and probably not to them, why they were heading for Halifax. News got around that there was a famous scientist on board, by name of Cockcroft. He was asked if there was some subject on which he could address the troops to keep them entertained. Cockcroft gave it some thought and decided to tell them about atomic energy. He gave a marvellous lecture in very elementary terms about radioactivity and the energy residing in the nucleus. He invented a new unit called the battleship-foot: the amount of energy necessary to lift a battleship one foot out of the water and, following a line of argument advanced by Rutherford, described how a single cup of water had enough energy to lift a battleship many feet into the air and probably break its back. This, he explained, was a very safe subject to talk about because there was no hope at all that such a thing would be achieved during the present war!

As we entered the harbour at Halifax, the first of the 50 US destroyers, which the US had offered Great Britain under lend-lease, also entered port. The reason for the large contingent of Naval personnel on the *Duchess of Richmond* was now clear. They would act as the crew for those timely gifts from the US. Waiting on shore was a US Army armoured vehicle, with submachine guns bristling from every orifice, together with an even bigger armed guard than the one at Liverpool. These had somehow been spirited across the Canadian border and were to transport the black box to Washington. As it was loaded on board I noticed that a neat pattern of holes had been drilled in each end.

At Halifax, Cockcroft and I separated from the rest of the Mission. They went direct to Washington to join up with Tizard and make preparations for our meetings with the US authorities. Tizard had insisted from the outset that Canada should be treated as an equal partner in any negotiations with the United States, and that the Canadians should learn everything which was disclosed to the Americans. On arrival of his flight in Newfoundland, Tizard had gone direct to Ottawa and had spent a whole week explaining the purpose of the Mission to the Canadian Chiefs of Staff and to the National Research Council. Only then did he go on to Washington to begin negotiations with the US authorities.

Cockcroft and I had the easy task of fixing the details with the National Research Council (NRC) in Ottawa. As a result, Professor Ralph Fowler, then acting as UK Scientific Representative in Canada, Professor Shenstone, a Canadian-born professor at Princeton University, and Brigadier Letson, an ex-professor of Physics, were invited to Washington to represent Canada at the meetings with the Americans.

There was another reason for going to Ottawa and, unfortunately, this misfired. Before leaving London, arrangements had been made for a recent model of a 200 Mc/s ASV radar to be shipped to the NRC at the same time as the black box. The intention was to install this in a Hudson aircraft of the Royal Canadian Air Force and to fly it to Washington as a demonstration of how far Britain had progressed with airborne radar. There were indeed several crates on the bridge of the *Duchess* addressed to the National Research Council, but it turned out that they were some other consignment which had no connection with the Mission. Unknown to Tizard and myself, someone in London had decided that the equipment would not be sent until the new Mark II ASV had been tested and was in production.

After a few days in Ottawa, I flew to Washington on the evening of 11 September, ready for the first round of discussions with the US Army and Navy which were due to start the following day.

The Tizard Mission in Washington

When he arrived in Washington, Tizard first established contact with Lord Lothian at the British Embassy. He found that office space at the Embassy was in short supply, so he set up a temporary office in the Shoreham Hotel. This was a short walk from the Embassy in beautiful surroundings overlooking Rock Creek Park and was the hotel at which members of the Mission stayed while in Washington. As it turned out, I remained in Washington for many months after the Mission departed and for most of the time I stayed at the Shoreham. I had a small bed-sitting room with bathroom and kitchen attached and at $100 a month I thought I was being robbed!

The Embassy could not spare secretarial help so Tizard appealed to his new-found friends in Canada. Dean Mackensie of the National Research Council kindly provided two secretaries from his own office— Doreen Geary and Vera King. They proved to be wonderfully efficient, well acquainted with security matters and most helpful in steering us through official procedures in Washington, which had subtle differences from those we were accustomed to in London.

Through Lord Lothian, Tizard made contact with the top levels of the US Army and Navy and arranged a series of meetings. Our main contacts on radar matters were with General Mauborgne, the Army's Chief Signal Officer, and Admiral Harold Bowen, the Director of the Naval Research Laboratory (NRL). Another important connection was with the National Defence Research Council (NDRC) which had just been set up under the chairmanship of Dr Vannevar Bush, with Karl Compton and Conant as his colleagues. The first round of meetings was scheduled to start on 12 September and these were held in the temporary Army and Navy headquarters on Constitution Avenue. In the first instance, they were devoted to describing new British developments and the way in which they had been used in recent operations. This went on for a week or ten days.

The usual form of a meeting was for Tizard to make some introductory remarks, which he always did in the simplest and clearest possible way. He then called on the appropriate expert to elaborate and produce documentary or filmed evidence, as the case may be. It was immediately obvious from the interest taken and the questions asked that we were faced with experts who either had first-hand experience of what we were talking about or had given serious thought to the same problems. Copies of the British documents and drawings were handed over to Army and Navy Intelligence who were responsible for distributing them within the Services and for overall security of the material. A complete report of the British radar disclosures was kept in note form by Colonel Mitchell of the Army Signal Corps and a similar report for the US Navy by Lieutenant Commander Goullet of the Naval Research Laboratory.

It was then America's turn and we were shown the radar systems which they had developed up to that time. This was done by visits to their principal research establishments—the Naval Research Laboratory at Anacostia and the Signal Corps Laboratory at Fort

Monmouth. I will not attempt to describe the US radar systems in any detail, but will summarise the Mission's findings in the following terms.

(1) We were agreeably surprised to find that both the US Army and Navy had progressed a good deal further in radar than we had been led to expect.

(2) Although they had been thinking of ship and aircraft detection from a much earlier date, it was not until mid-1936 that Page at the Naval Research Laboratory had succeeded in obtaining clear, pulsed echoes from an aircraft. This had led to the development of a 200 Mc/s shipboard radar, the CXAM, which was remarkably similar to the British CHL and on precisely the same frequency.

(3) Similarly, the US Signal Corps, starting a little behind the Navy, had developed a mobile air-warning radar, type SCR 270, and a searchlight-director radar, SCR 268, using 100 and 206 Mc/s respectively.

(4) The principal difference between the British and the United States position in September 1940 was that the British air-warning chain had been in continuous use as an operational system since 1938, while the American radars existed only in ones and twos and had seen little service use. The radar scientists at both US establishments had fought something of an uphill battle for recognition of their work and the need for it to go into service. By comparison, the British Government had recognised the need for air warning as early as December 1935, when £1 million had been allocated to provide air-warning coverage of the Thames estuary. In 1940, the complete British chain system had been in operation against the enemy for over a year and was proving a decisive factor in the Battle of Britain, which was being fought and won at that very moment. The Tizard Mission was thus able to convince the US Chiefs of Staff that radar was a supremely important weapon of war; once they were convinced of this, production orders for radars using the American design and for American equivalents of British radar equipment flowed in to industry in great profusion.

(5) The US had used radio altimeters for some time but had made no progress with airborne radar for the detection of ships and aircraft. This was a field in which Britain had a substantial lead.

(6) Lastly, in spite of many endeavours, the US had made little progress with high-power transmitters on waves of centimetre wavelength. Perhaps the greatest single contribution made by the Tizard Mission was to introduce the resonant magnetron to the USA. As we shall see, it was taken up with tremendous enthusiasm, as a result of which over a million magnetrons of every type were

produced in the USA. They were incorporated in hundreds of thousands of radar sets of every description and went into operation in every theatre of war.

During the first few weeks of the Mission the existence of the resonant magnetron had been hinted at but not described. The Army and Navy were told that we could generate substantial pulse power at centimetre wavelengths but the means of doing this had not been disclosed. At this point, the US authorities made a most important decision. They decided that so much needed to be done at metre wavelengths that the Army and Navy would continue to take prime responsibly for research and for the introduction of such radars into service, but that responsibility for more speculative work, like that on centimetre waves, would be vested in the newly formed Microwave Committee of the NDRC. This was a decision which had an immense impact on the subsequent development of radar in the USA. Tizard formalised these arrangements with the NDRC in private discussions with Vannevar Bush and Karl Compton on Friday 27 September, just before his departure from Washington. Cockcroft and I were then left to take it from there with Dr Alfred Loomis, Chairman of the Microwave Committee.

Our first meeting with Alfred Loomis had already taken place on the evening of 19 September in his apartment at the Wardman Park Hotel, only a stone's throw from the Mission headquarters at the Shoreham Hotel. Alfred Loomis, Karl Compton, Carrol Wilson, Cockcroft, Admiral Bowen of NRL and myself were present. We learned that the Microwave Committee, as a group, had just recently made a tour of industrial laboratories and research institutes where work on microwaves was taking place. This was exceedingly helpful to us in giving guidance on the laboratories we should visit in the next few weeks. Briefly, we learned that there was a good deal of work on transmitting and receiving tubes at wavelengths around 40 to 50 centimetres which appeared to be well in advance of British work. At wavelengths around 10 centimetres, the Bell Telephone Laboratory and GE Schenectady had much to offer in receiver technology, while MIT, Stanford and the Bell Laboratory were well advanced with waveguides and horns on this wavelength. However, there was nothing, even in embryo, which matched the British resonant magnetron. We quietly produced the magnetron and those present at the meeting were shaken to learn that it could produce a full 10 kilowatts of pulsed power at a wavelength of 10 centimetres.

The crucial meeting with Loomis took place at his home in Tuxedo Park over the weekend of 28 and 29 September. Crockcroft and I flew up from Washington to La Guardia on the Saturday morning and were met by one of Alfred Loomis's cars which took us to Tuxedo Park about 50 miles north of New York City. It was a beautiful time of year, and as a newcomer from Britain I was enormoulsy impressed by the beauty of the autumn colours in the elms, the oaks and the maples of the Catskills. Loomis's main home was a beautiful Colonial mansion within Tuxedo Park. Nearby was an older home which had been converted to a laboratory—the Loomis Laboratory—a tradition he had inherited from his grandfather who had set up the first such laboratory 50 years before. This was not all. Near the highest point of Tuxedo Park, he had built what he called his playhouse—an ultra-modern establishment with full air-conditioning, still a rarity in domestic homes in the USA in 1940, and automatic everything else, much of it of his own design.

Trained as a lawyer and investment banker, Alfred Loomis had an extraordinarily wide range of interests and excelled in all of them. Perhaps his greatest love was for science; he not only practised science but was a patron of science as well. Among his contempories he was aptly called '. . . the twentieth century Benjamin Franklin'. His early career showed an interesting parallel with both Cockcroft and Tizard, which may account for the rapport which was established between them. Like Cockcroft, Loomis served as an artillery officer in World War I. He then became the Chief of Research and Development at the Aberdeen Proving Ground where, by accurate timing methods, he brought about tremendous improvements in the technology of gunnery. It could be said that, by the application of scientific methods, he did for gunnery in the USA what Tizard did for military aircraft in Great Britain in World War I.

Loomis's interest in precise timekeeping continued after the war and in the late 1920s, before the advent of quartz as an accurate timing medium, he owned two of the most accurate pendulum clocks in the world at his laboratory at Tuxedo Park. Later, he became interested in ultrasonics and, in collaboration with R W Wood, was responsible for important advances in that field. Still another interest was in the electroencephalograph for the measurement of 'brain waves', a subject which he put on a sound instrumental basis. He was awarded a medal by the Franklin Institute for this work—one of the rare cases in which he allowed his part in such work to be acknowledged.

Loomis was a keen sailor. He often took part in the classic Newport to Bermuda yacht race and at one time built a J class yacht as a contender for the defence of the America's Cup by the New York Yacht Club. As an aside, it is interesting to note that many years later his son, Lee Loomis, continued this tradition and was the Chairman of the New York Yacht Club Committee which guided the defence of the Cup right up to, but not including, the year in which the Cup was

Figure 10.1 Dr Alfred Loomis, '. . . the twentieth century Benjamin Franklin', who orchestrated the revolutionary change from metre to centimetre-wave radar in the USA.

lost to some renegades from Australia.

During the 1930s, Alfred Loomis developed a close friendship with Ernest Lawrence and interested himself in the construction of Lawrence's cyclotrons in California. He played a crucial, if muted, part in Lawrence's search for funds and in the acquisition of the large quantities of special steel and copper which were required for the early machines. As if by magic, Alfred knew where to obtain the requisite materials. It was later found that he was either a shareholder in, or else had bought a controlling interest in, several of the companies concerned with these materials.

The extraordinary thing is the modesty with which all of this was done. I can clearly remember two separate occasions on which he told me how a 'certain person' had gone about getting a particular project started, some years earlier. Many years later, I discovered that the 'certain person' was none other than Alfred Loomis himself. He went to extraordinary lengths to avoid taking credit for any developments with which he was associated. So, when in 1940 he played a decisive part in converting US radar to microwaves, he took pains to see that his part in it was covered up.

At the Loomis Laboratory that afternoon, in addition to the small staff of the laboratory, were Carrol Wilson, Vannevar Bush's personal assistant at the NDRC, and Ed Bowles, the Professor of Electrical Engineering at MIT, a man destined to play an important part in the practical use of radar by the Armed Forces in the USA. We were given a field demonstration of a 10 centimetre Doppler radar system which could detect a light aircraft at a range of a couple of miles. It would not have won the Battle of Britain but it showed the way their minds were working. The equipment included Hansen horns and waveguides and a klystron manufactured by the Sperry Gyroscope Company. After dinner, we had a wide-ranging discussion of existing US Army and Navy detection systems and the improvements which could be made by using centimetre waves.

On Sunday 29 September, we were joined by Hugh Willis, a member of the Microwave Committee and Director of Research for the Sperry Gyroscope Company. His company had recently negotiated a contract with the Varian brothers at Stanford for manufacturing rights to the klystron. That evening we produced the magnetron once again, together with the drawings and construction details. The atmosphere was electric—they found it hard to believe that such a small device could produce so much power and that what lay on the table in front of us might prove to be the salvation of the Allied cause. It was agreed at once that the Microwave Committee would sponsor production of the magnetron and that the powerful Bell Telephone Company would be asked to undertake manufacture. Up to that moment, only twelve production models had been made in Britain.

I have no doubt that the New York telephone system did an unusual amount of business that night because by the time Cockcroft and I left Tuxedo Park the next morning the Bell Telephone Laboratory had been told that if the Boot and Randall magnetron came up to expectations, they would be given a contract by the Microwave Committee for a small production batch. We had first to check this with Tizard who was due to leave for home within a few days. We returned to Washington and Tizard agreed immediately. He was delighted with the way things were developing and encouraged us to push vigorously ahead.

In the next few days, Tizard wound up the business of the Mission, pleased with what had been accomplished in Washington. He made arrangements for his return to the United Kingdom by Clipper from New York to Bermuda and then via the Azores to London. He left Washington for La Guardia at 9.30 AM on 2 October. He especially asked that I should accompany him on that leg of his journey because he had several things on his mind. The trip from Washington to New

York was by DC3 and took about two hours. He had many things to talk about and I will try to summarise them.

(1) He was delighted by the reaction of the Microwave Committee and the decision to go for manufacture of the magnetron at the Bell Laboratory. 'Push it for all you are worth' he said.

(2) He was still fretful about the decision in London to defer dispatch of the ASV set to the USA. He had put a great deal of weight on the Mission arriving in Washington with an airborne radar ready to demonstrate. We had since ascertained that London was making amends. They were sending a Mark II ASV, a Mark IV AI and a Mark II IFF, and these were already on the way. He was particularly concerned that British promises to deliver did not slip a second time and he charged me with the responsibility of seeing that once these sets arrived in the USA, nothing should go amiss. He made it clear that I should stay on in the USA until this task was complete.

(3) He then went on to talk about his return to London in what I thought was an excessively pessimistic way. As he had predicted, the Battle of Britain was already being won, but he seemed very apprehensive about what he would find when he returned to London. This I simply did not understand at the time. At the conclusion of what must have been the most successful example of trans-Atlantic military collaboration which had ever taken place, he was in a strangely pessimistic mood.

Subsequent events showed that his forebodings were well founded. When he got back to London, the Battle of Britain was already won, but as one of the most important architects of that Battle, he found himself discarded. His job no longer existed and Whitehall no longer had a place for him to sit. Precisely the same fate shortly overtook the victor of the Battle of Britain, Air Chief Marshal 'Stuffy' Dowding himself, but that is another story. The British have mysterious ways of rewarding their great men.

11

The Resonant Magnetron and the Bell Laboratory

On 3 October, the day after Sir Henry's departure, I went to the Bell Telephone headquarters building at 463 West Street, New York City. After an opening discussion with the director, Mervin Kelly, and his deputy, Ralph Bown, I had a full morning's discussion with their powerful Valve Group—J G Wilson, Samuel and McNally. These were names which I knew from their published papers and from their short-wave transmitting valves, which had played such a valuable role in our early airborne radars. This was the first time I had met them in the flesh. Our discussion covered the design of their whole range of transmitting and receiving valves in the shorter wavelength range, but the focus of interest was on the British resonant magnetron, which they had already heard about from Loomis. I put the magnetron on the table and produced drawings of the internal construction. We had an animated discussion of the design details and I gave a rough indication of the requirements for running it, namely a magnetic field of 1500 gauss and a pulsed anode potential of 10000 volts.

They were fascinated by the design. If, at the start of the meeting, they had reservations about the extraordinary output which was claimed for it, they soon admitted that it had enormous potential. They saw no problems in reproducing it in their own laboratory right there in West Street. Most of the mechanical features of the design were familiar to them. The 'Housekeeper' seals for the filament and the output terminals were in everyday use, and they could reproduce the heavy duty oxide cathode. Even the gold seal, which was the final conjuring trick produced by GEC Wembley to convert Boot and Randall's laboratory model into a production model, was familiar to them.

Figure 11.1 An external view of the actual magnetron taken to North America by the Tizard Mission in August 1940. This is probably the only example of the original batch of 12 still in existence; it is now on display in the Museum of the National Research Council, Ottawa.

Not only did they see no difficulty in copying the magnetron, but they were most anxious to do so. Before doing this, it was desirable to run it up and we discussed arrangements for doing so. It was most convenient to do this, not at West Street, but in their Whippany Laboratory in New Jersey, where they already had a pulse generator of the requisite voltage and an electromagnet. We would have gone there at once, but I had commitments at MIT which would keep me occupied for the next few days, so we made arrangements to foregather at Whippany on the following Sunday morning, 6 October. I left the meeting with the feeling that here was a group which was not just interested in taking on another job, they were simply bursting to get at it. I finished a most productive morning by being entertained to lunch at a nearby Greenwich Village restaurant by that doyen of science writers—K K Darrow. He had been a revered figure from my youth for the wonderful summaries of various branches of physics which he wrote about so elegantly in the *Bell System Technical Journal*. He was a model which a budding science writer would do well to copy to this day. I cannot remember in detail what we talked about, but it was a very pleasant lunch and I felt a better man from just talking to such a person.

I had appointments with RCA at Rockefeller Center and at MIT which kept me occupied until Saturday afternoon. I then returned to New York. Ralph Bown had kindly invited me to stay overnight at his home in Summit, NJ. We spent a relaxed evening and for the first time I learned how to make a genuine American martini. We finished the evening by going to the local cinema where we saw Mickey Rooney in one of the Andy Hardy films.

The next day, Sunday 6 October, was to be a fateful day. Ralph Bown and I went to the Whippany Laboratories where Wilson, Samuel, and several of the Whippany staff were already gathered. They had brought the magnetron with them from West Street and they had dug up a large electromagnet of the kind that was common in university physics departments in those days. They also had a modulator which produced a pulse a few microseconds long at 10000 volts. We checked the magnetic field and ran the filament up to the required potential. We were conscious of the fact that the magnetron had last been run at Wembley two months before and that it had since been transported 3000 miles. Would it still work? Very gingerly, we switched on the anode potential and were immediately rewarded with a glow discharge about an inch long coming from the output terminal. I had not seen anything like this before and we looked at it in amazement.

Someone said 'And what wavelength do you think that is?' I murmured that it should be around 10 centimetres. A wavemeter was produced and he said 'Dead right—9.8 centimetres. And how much power do you think it is giving?' I had been quoting the figure of 10 kilowatts given to me by Megaw, but looking at that glow discharge I said 'That must be much higher than 10 kilowatts, say 15 kilowatts.' One of the Whippany staff pointed to a large 10 foot cubicle in the corner of the room which turned out to be a 500 Mc/s pulse transmitter for the Navy. 'How much power do you think that gives?' Then, ruefully, 'About a kilowatt.' As a demonstration it could not have been more successful. Some decisive talks must have taken place between Loomis and Mervin Kelly in the previous few days because the Valve Department had already received their instructions. If the magnetron worked anywhere near its promise, they were to commence manufacture of 30 copies for the NDRC immediately. We went away from the laboratory walking on air and I returned to Washington well satisfied that the magnetron had been planted in fertile ground.

But the following day I got a rude shock. I received an urgent telephone call from Mervin Kelly in New York. He was quite agitated. 'Could you come up to New York—right now or tomorrow morning at the very latest.' Over an open telephone line he could not be more specific, but clearly something drastic was wrong. 'Oh my God,' I thought 'the magnetron must have blown up.' I was up at 5.30 next morning and caught a 6.45 flight to New York. I went straight to West Street and was greeted with great restraint in the foyer by someone who had obviously been instructed to look out for me and conduct me the Conference Room on the top floor. I did not have a clue what to expect and awaited developments with some apprehension. This was not the warm greeting I had received only a few days before. Mervin Kelly came into the Conference Room with Wilson and some of the Valve Laboratory staff, together with two or three other gentlemen whom I had not previously met. They looked suspiciously like company lawyers, and, again, I wondered what was in the wind. Mervin Kelly greeted me in the same restrained fashion and we took our seats around the conference table. Kelly acted as chairman and introduced the subject by outlining the events of the past few weeks—the arrival of the Tizard Mission, its purpose, the delivery of the magnetron to the Bell Laboratory and the test run at Whippany on 6 October. He still gave no clue as to what the meeting was about. Wilson then produced the magnetron and laid it in the middle of the table. In rather sombre tones Mervin Kelly said 'Is that the magnetron that you brought here last week?' I agreed that it was. Wilson then produced the blueprint that I had given them, with the internal dimensions quoted to a hundredth of an inch. 'And is that a diagram of the interior of that magnetron?' I had to agree that it was the same drawing, with the GEC Wembley imprint on it. There was a long pause, and Mervin Kelly then produced a photograph from a drawer at his end of the table. 'It may interest you to know' he said, pointing dramatically at the magnetron lying on the table, 'that this is an x-ray picture of that magnetron'. I looked at the photograph and could not believe my eyes. Every account I knew about, from the first sketchy descriptions of the Birmingham model to the detailed accounts of machining and assembly which I had seen at Wembley, indicated the resonant system had six holes. The x-ray picture showed eight holes!

There was little I could say and I asked if I could get in touch with Megaw immediately, by transatlantic telephone or cable. I could not have been in better hands and it was not long before I was talking to Megaw. I told him what had happened and wondered if he could explain it. At first he was vague and disbelieving—he was obviously as puzzled as I was. Then the penny dropped and he said 'Oh my God, now I remember.' Then the story tumbled out. He had placed the order for the first twelve sealed-off magnetrons in the experimental shop at Wembley late one afternoon towards the end of July. Later that night, having thought about it for a while, he went back to the shop and told the foreman 'Look, on that order for twelve magnetrons, make ten with six holes, one with seven and one with eight.' Under considerable stress, as we all were at that time, he had then gone away and forgotten all about it. The jigsaw puzzle rapidly sorted itself out. We both remembered that in the batch we had checked at Wembley, there was one which did not work. He still had it on hand and he was able to confirm by cable a few days later that the dud magnetron was indeed Number 11 and it had seven holes!

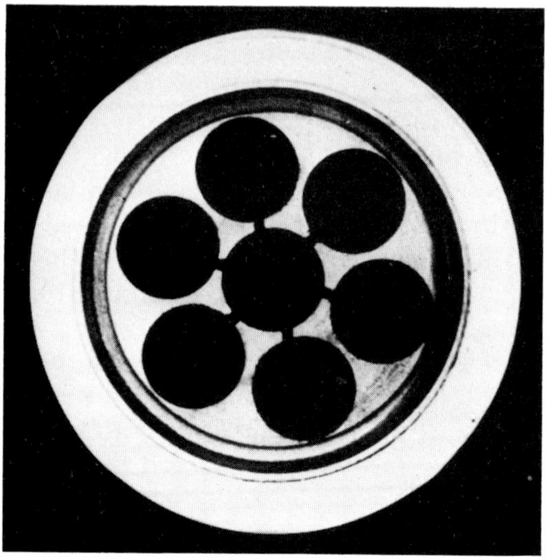

Figure 11.2 The core of the original resonant magnetron, invented by Boot and Randall at Birmingham University in February 1940. The first 12 production models, Type No E1189, were produced at the GEC Research Laboratories, Wembley, in July 1940.

So the whole episode ended happily, but there was still an important decision to be made. Should the Bell Laboratory copy the GEC drawing or should they copy the x-ray picture? We agreed that there was only one way to go—to copy the one which they knew worked. That is the story of how all the early British magnetrons had six holes and all the American ones had eight.

The rest of the story is quickly told. The first magnetron was working in the Bell Laboratory within a few weeks and all 30 were produced in a little over a month. The first five went to the Radiation Laboratory, which was just being formed at MIT. The Radiation Laboratory also set up its own magnetron group which, like the one at the Bell Laboratory, produced a whole range of magnetrons operating down to millimetre waves and giving powers which were to reach one megawatt before the end of the war. The demand was so great that many other manufacturers were brought in, among which GE, Westinghouse, RCA and Raytheon were prominent. Counting all the different types, well over a million such magnetrons were produced before the end of the war, and they were built into hundreds of thousands of radars of different types. This all came about from the one magnetron type E1189 from the GEC Research Laboratory in Wembley.

Interspersed with these visits, members of the Tizard Mission made

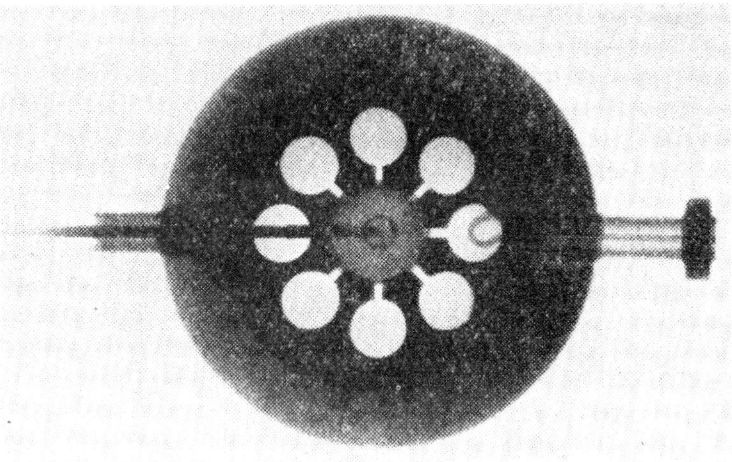

Figure 11.3 An x-ray picture of Number 12 of the first production batch. As described in the text, this had 8 holes not 6. (Courtesy of Bell Telephone Laboratories.)

a round of visits to research establishments, most of them on the east coast, where both centimetre-wave and decimetre-wave work was in progress. In this, we were guided by the previous tour made by Loomis's committee. These visits were made by individual members of the Tizard Mission acting alone, or in small groups of two or three, with Cockcroft, Brigadier Wallace and myself playing the principal roles. In short, we found that the Americans were substantially ahead in centimetre-wave receiver design, in crystal mixers and in components like waveguides and horns, but that the US had nothing to offer in pulsed transmitters which could compare with the British magnetron. At most of these laboratories we were able to pick up samples—always freely given—of klystrons, acorns, lighthouse tubes, diodes of every description and component parts likely to be helpful to our colleagues back in Britain. These were carefully packed in Washington, together with descriptive matter, and dispatched to London in the diplomatic bag. Each visit was carefully reported upon, often by more than one member of the team, and these reports were dispatched to London and Ottawa for further distribution. Covering all subjects, a total of about 140 such reports were written and a complete record exists to this day under Catalogue No AVIA 10/1 to 4 at the Public Records Office in Kew.

Among the above reports are many about visits to the other Bell Laboratory at Holmdel, NJ, where Harald Friis, a particularly charming man of Danish extraction, led an exceedingly talented team. This included Southworth, the waveguide specialist, Karl

Jansky, who a few years before had laid the foundations of the new science of radio astronomy, and the three engineers, Englund, Crawford and Mumford, who had been the first to publish a report about observation of radio reflections from an aircraft in 1933.

The Holmdel Laboratory was principally concerned with microwave receivers and antennae, and with microwave propagation, which they had been engaged upon for several years. Friis had 10 centimetre mixers of exquisite design and a whole range of signal generators, wavemeters and other measuring devices which made my mouth water. Friis himself was a meticulous experimenter and I learned a great deal about experimental procedures just by watching him at work. These men were masters of that particular field.

12

Formation of the Radiation Laboratory

With production of the magnetron under way, Loomis began to think about implementing the responsibilities the Microwave Committee had taken on for research and development of centimetre-wave radar systems. Clearly, a new research establishment was called for and there were many important questions to be answered. Where should it be established, who should staff it and which were the most important projects to be undertaken?

This was the main subject for discussion at the next visit of Cockcroft and myself to Tuxedo Park, which took place on the weekend of 12 and 13 October, 1940. This time, Professor Ernest Lawrence was present at Loomis's invitation, together with Ed Bowles from MIT and Carrol Wilson, representing NDRC. We first discussed the most urgent projects which should be tackled. Loomis paid us the compliment of indicating that these should be the projects most needed by Britain at that moment. It did not require much discussion to identify them:

(1) a 10 centimetre airborne radar, with air interception as the first objective;
(2) a 10 centimetre gun-laying (GL) radar for anti-aircraft gun-laying;
(3) a long-range navigation system, of a kind not yet specified.

It is rather easy to identify these as the special interests of Tizard, Cockcroft and myself (in reverse order). Tizard had long since insisted that North America, with its great distances, was the ideal place to work on a long-range navigation system. Testing in Great Britain was severely limited because there was literally no place to go to check on the accuracy of a long-range system without being shot at by hostile aircraft. For his part, Cockcroft had already been thinking about a successor to the British GL radar Mark I and II, which operated on a much longer wavelength and had some severe limitations. On airborne radar, I had long been advocating centimetre waves as the next direction in which to go, both for air interception and ship detection.

Figure 12.1 Outside Alfred Loomis's house at Tuxedo Park, 12 October 1940. This was the second meeting at Tuxedo Park, when plans were laid for the formation of the Radiation Laboratory at MIT. From right to left: Alfred Loomis, Professor Ernest Lawrence, 'Taffy' Bowen, Professor Edward Bowles, Frank Lewis and Carrol Wilson. The photograph was taken by Professor John Cockcroft, the only other member of the Tizard Mission present on that occasion.

The discussion was animated, but we did not work too hard. After all, it was a Saturday. I remember adjourning for tennis and being thoroughly beaten by a very expert Ernest Lawrence. Before dinner that night, Averil Harriman, who was a neighbour of the Loomis's, dropped in to say hello and to have a quiet drink. It was a very different world from the one in which I had grown up, in the Swansea Valley only a few years before.

The next morning, we discussed what form the new laboratory should take. Loomis quickly agreed that it should be a civilian establishment modelled on AMRE and ADRDE, with close links with the military and with industry. Both would be easy. The Army and Navy had already indicated they would welcome a contribution from NDRC on centimetre waves; and, except for Loomis and Bowles, the membership of the Microwave Committee already consisted entirely of representatives from industry. Several localities were mentioned as suitable for setting up the laboratory, mostly in the Washington area, but no decision was taken. Then the question of staff came up, and here there was a serious problem. Where should they come from? Clearly they could not come from industry, which was already beginning to recruit people for radar at a high rate. Cockcroft and I described how university people had been drafted into radar establishments in Britain—not only physicists and engineers but scientists from practically every discipline—and they were taking to military research like ducks to water. The point was made. Ernest Lawrence was particularly enthusiastic about recruiting from the universities.

That point settled, we got back to the three main projects. I undertook to write a specification for an airborne radar, and Cockcroft a corresponding specification for a gun-laying radar. Long-range navigation was thrown open for discussion. It was generally agreed that the range should be 1000 miles or more and the accuracy should be about ± 5 miles. No one, at that stage, came up with a specific proposal, but I described Dippy's Bawdsey proposal of 1938 for a system which later became known as GEE, the range of which was limited to a couple of hundred miles.

Loomis then proposed that, since the magnetron was under way, the first task of the new laboratory was to assemble a 10 centimetre radar system which could be tested against echoes from ground objects in the manner which was customary in short-wave systems. We listed desirable figures for things like the pulsewidth, receiver sensitivity, cathode-ray tube display and so on. On scratch pads and the backs of envelopes, we sketched the block diagram of a typical system right there, with a modulator, a transmitter incorporating the magnetron, a receiver and indicator and appropriate power supplies. Loomis surprised us by saying that he would call a meeting of the Microwave Committee the very next day and settle the details. We drove down to New York City and gathered in his apartment at 21 East 79th Street about 11 AM the next morning. Unfortunately, Cockcroft had another commitment and could not be present and, due to the short notice, the members from GE and Westinghouse could not attend. A record of the meeting does not seem to exist, but a note in my diary gives the names of those present and the items discussed. This is given in Appendix 5. Loomis outlined his proposal of the previous day and this visibly shook those who had not been at Tuxedo Park. He further astonished them by asking the representatives from industry to select which parts of the project they wished to tender for and to attempt to deliver their contribution in 30 days. They indicated that there would be no problem submitting tenders in 30 days. 'No' said Loomis 'you misunderstand me. I want you to submit your tender next week and to deliver 30 days after that.' His audience was stunned, but they deferred judgement until they could get back to their respective laboratories and discuss it with their colleagues.

When pressed to express opinions, the Bell Laboratory confirmed their interest in supplying the magnetron—it was already under way. GE wanted to make the magnet, RCA the pulse modulator, Sperry the paraboloid and scanning gear, and RCA the power supply. BTL, RCA and Sperry all expressed interest in tendering for the receivers and indicators, and RCA in a special delay screen cathode-ray tube based on the British model. This was Monday 14 October and the next meeting was arranged for Friday 18 October in Washington! Only five weeks had passed since the arrival of the Mission in Halifax and things were moving exceptionally fast.

The Committee broke up, but Loomis (who must have been working a 24 hour day) had two further things to discuss with Ernest Lawrence and myself. Overnight, an idea had come to him about long-range navigation. What about a pulsed hyperbolic system, like GEE, but on long waves which would be reflected from the ionosphere and would, therefore, give a range of one or two thousand miles. Since the two ground stations would themselves be about a thousand miles apart, there was a problem of synchronisation. This he proposed solving by using his specialty—in this case highly accurate quartz clocks—at each station. It sounded like a marvellous idea and, right there, Project 3 became quite specific. This was the birth of LORAN, a navigational system which has survived to this day.

The second point he wanted to raise was the question of recruitment. Would Ernest Lawrence be prepared to act as the principal recruiting agent? Anyone who knew Ernest Lawrence in those days will remember him as a man of sudden enthusiasms—and when he became enthusiastic about something, he went after it like a dynamo. After 45 years, I cannot be exactly sure about dates and times, but during the next few weeks he was to telephone every physicist of consequence in the United States. He told them that something very extraordinary was about to happen and their services would probably be needed somewhere on the east coast, if not within a week then certainly within a month. He began using the telephone right there in Alfred Loomis's apartment and the telephone bill must have been astronomical.

The next meeting of the Microwave Committee, on 18 October, was little short of epoch-making. It was held in the Carnegie Institution, on P Street in Washington DC, of which Vannevar Bush was President and Alfred Loomis a Trustee. This was the venue from which Van Bush and Karl Compton steered the destinies of NDRC (later the Office of Scientific Research and Development, OSRD) until the end of the war. The only record of this meeting which survives is another note in my diary (Appendix 5). A list of those present reads like a meeting of the top four hundred. Loomis as Chairman, Vannevar Bush, Karl Compton, Dr Coolidge, the revered head of the GE Research Laboratory at Schenectady, and Dr Jolliffe, chief of research at RCA. The only member of the Committee proper who was missing was Mervin Kelly of the Bell Laboratory, who was unavoidably committed elsewhere. The US Army and Navy were represented by Colonel Mitchell and Major Smith from the Signal Corps, and Commanders Briscoe and Tucker represented Naval Research.

Loomis outlined the events of the previous few weeks, indicated NDRC's decision to go ahead with centimetric radar and its plan to set up an independent laboratory to carry this out. He then came down to details of the tendering arrangements. The Bell Telephone Laboratory set the pace and said they were quite sure they could deliver 5 magnetrons in 30 days. GE gave a similar assurance for a permanent magnet, which they estimated would weigh 2.5 pounds and provide a field of at least 1500 gauss. Sperry promised the scanner within 30 days. Bendix, as aviation experts, had come in as contractor for the prime piece of equipment—the power supply which would ultimately be required to run off an aircraft engine. Westinghouse promised a modulator in 30 days but RCA were hesitant; they wanted 90 days, which considering all the circumstances was not unreasonable. On receivers, the Bell Telephone Laboratory promised 5 crystal mixers in 30 days, Sperry the same, with the klystron oscillators perhaps a week later. RCA were again conservative and promised two 10 centimetre receivers in 90 days; but on cathode-ray tubes with delay screens they were able to provide a shorter delivery—between 2 and 4 tubes in 30 days. Centimetric radar was on its way in the USA. It was a momentous meeting and the consequences are with us to this day. Most of the promises made at that meeting, extravagent as they might have appeared at the time, were in fact met. What it meant was that roughly a month later there became available at MIT virtually a complete 10 centimetre laboratory radar system of exceptional performance.

I have already mentioned Loomis's plan to form a special NDRC

Laboratory to carry the project through. The first locality suggested was the Department of Terrestial Magnetism in Washington (DTM), already closely associated with the Carnegie Institution. However, when Loomis enquired into this possibility, he found that another NDRC group was already being formed at DTM to develop the proximity fuse. The next choice was Bolling Field, an Army Air Corps base on the outskirts of Washington. It soon became apparent that Bolling Field was not a practical proposition. The Army Air Corps wanted to move out but they knew that an alternative site would not become available to them for several years.

However, there was now another strong contender in the field. Vannevar Bush was the ex-Dean of Engineering at MIT. Karl Compton was the President of MIT and Alfred Loomis a Member of the Corporation. The answer was obvious. With all the scientific talent which Lawrence was recruiting shortly to come together in one place, what could be better than MIT itself. Karl Compton proposed that MIT should make 10000 square feet of laboratory space available within the Main Building. If aircraft facilities were required, MIT had access to a 160 × 130 foot hangar at Boston Airport, with a modern machine shop attached. This proposal was irresistible and was adopted by NDRC forthwith. I have no idea what formal arrangements were made in this connection, but judging by Loomis's prior performance they would be minimal. After all, he had just committed himself to serve the US Government for the princely sum of $1 per annum.

Meetings of Loomis's Microwave Committee were now held at almost weekly intervals—Wednesday 30 October at MIT itself, Thursday 7 November in Loomis's apartment in New York City and Friday 15 November at the Carnegie Institution in Washington. At the 30 October meeting, Lee DuBridge was already present. He was among the first to respond to Lawrence's call and it was assumed from the outset that he would be Director of the laboratory. He was a highly respected and very talented Professor of Physics from Rochester and, in my experience, was at once the best qualified and the most pleasant scientist to lead a radar laboratory during the whole war. He handled technical meetings with great aplomb and I never heard him express an angry word. On the rare occasions when proceedings showed signs of getting a little tense, he was always the one to keep a cool head and smooth a ruffled situation.

At the MIT meeting, Committee members reported optimistically on progress in the manufacture of components for the first centimetric set. There were no delays and they all expected to be on time. Space was rapidly being cleared within MIT and it looked as if laboratory work could start within weeks. We examined the space

Figure 12.2 Dr Lee DuBridge, Director of the Radiation Laboratory, and Professor I I Rabi discuss the resonant magnetron with 'Taffy' Bowen at the Radiation Laboratory, MIT, in 1940.

becoming available and it was clearly enough to give the laboratory a flying start. It was all on the ground floor of the Main Building of MIT and, remembering the virtues of roof space at Bawdsey, I hinted at the need for access to the roof. The response was immediate and in no time at all a wooden hut about 50 by 20 feet was built on the roof. This was to become a hive of industry and something of a focal point, where the first complete set came together to be tested against ground targets in Boston, just across the Charles River.

By mid-November things had really begun to move. After the 15 November meeting in Washington, I travelled with Alfred Loomis, first to Tuxedo Park where Ernest Lawrence had virtually completed his recruiting campaign. Loomis, Lawrence and I then went by car to MIT to inspect progress at what was now known as the Radiation Laboratory. The transformation there was remarkable. The magnetrons had arrived from Bell Laboratories and the other components were beginning to appear. Lee DuBridge was hard at work as Director, supported by a veritable galaxy of talented people from all over the US. I I Rabi, Ken Bainbridge, Lou Turner, Eastham, Alex Allen, Ed Macmillan, Luis Alvarez, Barrow and Hall were already installed, with many more to follow. In later years, six or seven of the early recruits became Nobel Laureates, not for what they did at the Radiation Laboratory, but for their expertise in entirely different fields. Many of them moved on to play crucial roles in the Manhattan Project, but that is an entirely different story.

A loose structure was beginning to form, with Rabi in charge of magnetron research, Turner in charge of modulators, Allen and Hall responsible for scanning antennae, Barrow for waveguides and horns and so forth. A small problem emerged in that nothing had been done about a gas discharge tube for common aerial operation. The Naval Research Laboratory at Anacostia had long since solved this problem at longer wavelengths, so the matter was quickly put right in consultation with them, and by the placing of an additional contract with RCA for a gas discharge tube modified to operate within a waveguide. The first complete 10 centimetre system came together during the last two weeks of December 1940 and there was every expectation that it would be working and receiving echoes early in January.

Installation of the First Airborne Radars in the USA

While the Radiation Laboratory was being formed, I still had a major task to peform, which had been delegated to me by Tizard—the installation of 200 Mc/s sea-search and air-interception radars in US aircraft. This was a particularly welcome undertaking; the sets were the very radars which I had developed over the previous four years in Britain. There has often been argument about which country first built or operated certain types of radar equipment and these arguments will no doubt continue well into the future. However, there is little doubt that Britain was well ahead of most other countries in the design of airborne radar for air interception and ship detection.

It will be recalled that the original plan was to bring an ASV Mark I with us in the *Duchess of Richmond*, install it in a Hudson aircraft of the Royal Canadian Air Force (RCAF) at Ottawa and fly it down to Washington for flight demonstrations. Unknown to us, this plan was thwarted somewhere in London, when it was decided that the shipment would be delayed until the Mark II ASV radar was in production. During the first few weeks of the Mission, this led to some rather acrimonious cabling between Washington and London; by the time of Tizard's departure from America in October, we knew that the situation had been retrieved in a handsome fashion. Not one, but all three versions of our airborne radars—a Mark II ASV, an AI Mark IV and an IFF Mark II—were on the way to the USA by surface ship. These were consigned to NRC in Ottawa, as previously planned. A revised programme was therefore drawn up to install the ASV radar in a Consolidated Aircraft PBY of the US Navy at Washington and to fit the AI radar to an Air Corps Douglas A20 at Wright Field in Dayton, Ohio. I made arrangements for the loan of these aircraft with Air Commodore Baker of the British Air Commission in New York on the afternoon of 30 September, only a few days before Tizard's final departure from America.

Air to Surface Vessel Mark II

The airborne radar shipment duly arrived at Ottawa towards the end of October and I was there to check it before onward shipment to Washington. The RCAF had already expressed interest in manufacturing copies of the ASV Mark II in Canada, and both NRC and the RCAF offered to help with the installation in Washington. This was too good to miss and I gratefully accepted. Heath, an experienced radio engineer from NRC, and Squadron Leader Williams of the RCAF were assigned to help. They arrived in Washington with the equipment on Friday 1 November and went direct to the Naval Research Laboratory. The plan was first to run the ASV equipment in the laboratory, and then install it in the PBY which was already available at the adjacent Naval Air Station.

When delivered, the ASV radar included a regular set of forward-looking antennae, but Heath and Williams decided to be ambitious and install some sideways-looking aerials as well. With the enthusiastic help of the Radio Division at Anacostia and the installation engineers at the Naval Air Station, this was accomplished in a remarkably short time. Heath and Williams made a short test flight around Washington during the last few days of November 1940 and, although they did not go out to sea, they reported strong ground signals out to a range of 50 miles. Everything appeared to be functioning well.

I took my first flight with them on Monday 2 December 1940, when we went well out into the Atlantic and checked the performance against both small and large merchant ships. The performance was clearly better than our old ASV Mark I and we received one echo from what must have been a capital ship in the Atlantic at a range of 60 miles. Strangely enough, my last ASV radar flight had been exactly a year before—on 2 December 1939 with the older Mark I ASV—and the improvement was very obvious.

A photograph of the PBY showing the long-range receiving antenna is given in figure 12.3. During the next few weeks, demonstration flights were given to a number of US Navy officers—to Admiral Noyes, Commanders Furth, Owen and Cleaves of the Navy Department—and several officers of the British Air Commission who had not previously flown with ASV equipment.

Figure 12.3 The first airborne radar fitted to an American aircraft. This was a British ASV Mark II, fitted to a US Navy PBY at the Naval Air Station, Anacostia, in December 1940. The forward-transmitting antenna is visible in the nose of the aircraft and one of the sideways-looking receiving arrays can be seen on the port side of the fuselage.

The whole operation was voted a great success and the US had been introduced to airborne radar for the first time. The Navy Department took over from that point and shortly placed orders for no less than 7000 copies of ASV Mark II from the Philco Corporation. In addition, a further 10000 sets were manufactured by Research Enterprises Ltd in Toronto, the majority of which were used by US Forces. They were put into operation over the Atlantic, in the Pacific and in practically every other theatre of war.

Identification of Friend from Foe

Even before the ASV had been put into operation in a US aircraft, the IFF Mark II, which had accompanied the shipment from Ottawa, was installed in a US Navy fighter by NRL personnel and test flown against the Naval Research Laboratory's CXAM radar. Once again, it was a demonstration which worked the first time, when it achieved a range of 90 miles. I remember discussing this with Hoyt Taylor on 20 November. He understood the vital importance of a common method of radar identification between British and US Forces. He was satisfied that this was the equipment to use and arrangements were made for both the US Navy and the Army Air Corps to adopt IFF Mark II as their standard radar identification equipment. Procurement, which finally ran into tens, if not hundreds, of thousands of sets was instituted.

Air-Interception Radar Mark IV

At about the time the PBY installation was being made, a Douglas A20 was also fitted with an AI Mark IV at Wright Field. I visited there on 10 October to make the detailed plans. The AI Mark IV arrived from Ottawa during November but I was so heavily involved with the ASV radar in Washington and the beginnings of the Radiation Laboratory that I could not supervise the Wright Field installation. This was handled by Squadron Leader Hignett of the British Air Commission and this was one of the rare occasions on which I did not manage to take a test flight and check on the details of what was done. This was twelve months before the entry of the USA into the war and night interception was not regarded as an urgent problem. Nevertheless, a small production order was placed with the Western Electric Company for Mark IV AI radar under the type number SCR 540.

By the end of 1940, the first airborne radars had been successfully installed in US aircraft and the Radiation Laboratory at MIT was thriving. So, I went off with Loomis for a well earned Christmas break at Hilton Head Island. This was his South Carolina hideaway, where the whole Loomis family usually gathered for a Christmas holiday.

Growth of the Radiation Laboratory

Right from its inception, an outstanding feature of the Radiation Laboratory was the camaraderie which prevailed. In many ways it was more like a scientific convention than a research laboratory, except that it was a convention which kept running year after year. Everyone worked long hours and did not spare themselves. Here was the cream of American scientists, hell-bent on doing all they could for the war effort—some fourteen months before America itself actually entered the war. If my analogy is correct, the Radiation Laboratory must rank as the longest running scientific convention of all time.

If there was any relaxation, it was on Friday night—in the bar of the Commander Hotel just behind Harvard Square. Inevitably, this became known as Project 4 and, although attendance was not compulsory, there was always a good turn-out and the hotel did a roaring trade. I remember it particularly well because there was a mural around the walls which depicted various episodes dear to the heart of all Americans—the Boston Tea Party, Paul Revere's Ride and Minute Men, dressed like characters out of Fenimore Cooper, shooting through the trees at highly visible Redcoats. The message was loud and clear—this was where they beat the pants off the British. Proceedings usually began with an expression of solidarity, a friendly toast to the British—'The hell with the Limeys.' It was all in good humour and there were many friendly barbs about certain failings of British tactics in the battles which finally lost them a big slice of the North American continent.

By mid-November, the recruitment rate at the Radiation Laboratory had reached one eminent scientist a day. They were invited to come to Cambridge, ostensibly to find out what was going on. They were often so excited by what they found that many of them did not go back to their home towns. The called their wives, told them to rent the home, pack a few bags and come on to Cambridge. Many put up temporarily in the Commander Hotel and then quickly settled in surrounding townships with evocative names like Concord, Bedford and Lexington, where housing was relatively easy to come by.

By the early part of 1941, under Lee DuBridge's wise guidance, the staff arrangements had settled down to something like their final form. Lee DuBridge was Director, with Rabi and Wheeler Loomis as Deputy Directors. Rabi was a fabulous character, born in the Bronx of German–Jewish parents, with an inbuilt hatred of the Nazi regime. He was one of the supreme architects of the radar revolution in the US and he later played a significant part in the production of the atom bomb. When the Radiation Laboratory got under way, he was the man principally concerned with scientific matters, while Wheeler Loomis looked after personnel and further recruitment. About this time, several more distinguished names began to appear—Ed Purcell, Zacharias, Bob Bacher, who took charge of Indicators, Louis Ridenour, who guided the gun-laying project and Lawson, who took charge of advanced development in the Roof Laboratory.

On 4 January, the roof system picked up its first echoes from buildings in Boston across the Charles River basin. This was still a two-antenna operation, with separate paraboloids for transmitter and receiver. There was a short and rather frustrating period when trouble was experienced with the design of the duplexer which permitted single-antenna operation. For the first, and possibly the only, time a mood of pessimism crept into the group and some doubts were even expressed about whether the system would ever be capable of receiving echoes from an aircraft. However, with the duplexer properly tuned and lined up, success was achieved on 7 February when the first echoes were obtained from aircraft flying over Boston Airport at a range of four to five miles. This was just in time to be reported at a meeting of the Microwave Committee which met that day in Washington and there was general agreement that the project was well and truly launched.

From this point, the Radiation Laboratory surged ahead; the staff continued to increase in number and the budget swelled. I remember a temporary hitch did occur in the summer of 1941. Most of the staff had been given one year appointments and the universities from which they had come were anxious to know whether they would return the following year. America was not yet in the war, but there could be no doubt about the success of the Laboratory and everyone in authority was confident that it would continue. NDRC had already asked for finance for the second year but Congress was slow in giving approval. A sum of one million dollars was involved for salaries alone. The Corporation of MIT voted to give half this amount from its own funds, but there was still a shortfall of half a million dollars. Representations were made to John D Rockefeller Jr, who immediately agreed to underwrite that sum. I am sure that the fact that Karl Compton and Alfred Loomis were behind the request had much to do with this. Needless to say, with the threat of war looming on the horizon, Congress was persuaded of the importance of going ahead, and both the MIT Corporation and Rockefeller were soon repaid.

Returning now to the technical story, let us take each of the projects in turn.

Project 1: Airborne Radar

No real attempt had been made to engineer the first roof set in a form which could be installed in an aircraft, although the sizes and weights were well within reason—that is, hundreds of pounds per unit instead of the tons we had started with five years before at Bawdsey. The first priority in 1941 was to re-engineer each of the 10 centimetre-wave units in a form suitable for aircraft installation; this was usually done by construction of new units within the Radiation Laboratory, in workshops which were exceedingly well equipped.

A Douglas B18 had been assigned to the project by the Army Air

Corps and this was accommodated in the National Guard Hangar at Boston's Logan Field. The hangar was later taken over by the Radiation Laboratory as its main Flight Facility. Over the years, the flight grew to include many aircraft of both the Army Air Corps and the Navy, with a sprinkling of civilian machines for routine tasks. Ed Macmillan took charge of the prototype airborne unit and began installing it in the B18 towards the end of February. Ground tests were completed early in March and the equipment was declared ready for the first flight. This took place on 10 March 1940, when good echoes were received from objects on the ground. Quite fortuitously, it happened that this was the very day that my colleagues back at Christchurch made their first air-to-air test with the British 10 centimetre equipment. On that occasion, they were successful in getting echoes from another aircraft at a range of one or two miles. The flight at Boston was only against ground targets and it was a few weeks before air-to-air signals were received.

During the next couple of weeks, the performance of the set was gradually improved and the first flight using an aircraft as a target took place on 27 March with Ed Macmillan, Luis Alvarez, Ernie Pollard and myself on board. It was to be an epoch-making flight. The target was a single-engined machine borrowed from the National Guard and, to get clear air, we flew out over Cape Cod Bay at about 10000 feet. We did several runs on the target and got satisfactory ranges of detection of two to three miles. We were delighted to see that this was not limited by the ground return—a prime reason for going to centimetre waves in the first place.

We were very pleased and might have concluded the flight at that point. However, there were several merchant ships in the Bay, so we switched off the elevation scan, left the set giving range and azimuth signals only and did a run on a ship of some 10000 tons in the Bay. We do not have a record of the maximum range at which it was detected, but it was about 10 miles. This was not exceptional compared with the performance of a standard 200 Mc/s ASV radar of that time, but for the first flight of a centimetre-wave radar it was a great performance. My colleagues from the Radiation Laboratory were wildly excited and could not contain their enthusiasm.

This result was so good that I suddenly had a thought. 'Let's go down to New London and see if we can find a submarine.' New London, in Connecticut, was an important US Navy Submarine Base and one of the main manufacturing centres for submarines in the USA. It could be reached in about 30 minutes flying time, so down we went. We were lucky enought to find a large submarine cruising off-shore; it was fully surfaced and presented an excellent target. We did some runs broadside-on and obtained a maximum range of four to five miles—another dramatic result.

We returned in triumph and the news spread around the Laboratory like wildfire. From that moment, ship and submarine detection was added to the project list and received just as much attention as air-to-air work. The results against a submarine were of particular interest to the US Navy. German submarines had already begun to appear in American waters close to the US east coast and were harassing the vital trans-Atlantic freight route. Although America was not yet in the war, they realised that here was a method of dealing with the menace when the time came.

The date 29 April 1941 was something of a red-letter day. We received a visit from Air Chief Marshal Sir Hugh Dowding, the victor of the Battle of Britain and now on a tour of inspection of US military establishments. From his earliest days at Fighter Command he had always been keen to see research for himself; he asked what was up and coming at the Radiation Laboratory. He was surprised to find that things were so advanced and even more surprised when he was taken for a flight in the B18. To me, this was reminiscent of the flight at Martlesham just two years before, when I had shown him the first complete 200 Mc/s air-interception set in a Fairey Battle. Lee DuBridge accompanied us on this flight and with Ed Macmillan in charge of the gear we gave the usual demonstration against an air target. By this time the maximum range of detection had grown to 5 or 6 miles at a height of 10000 feet—a very satisfactory improvement over a conventional 200 Mc/s set. Sir Hugh was still interested in the minimum range performance and repeated his request of two years before. Would we please go down to minimum range and let him judge for himself whether this was close enough. He watched the range close to a minimum on the cathode-ray tube and when this was reached he asked the pilot to hold it. He came up from below—the installation was in the nose compartment below the main cockpit—and found that we were practically alongside the target, actually at something less than 500 feet. He declared himself quite satisfied and, in my experience, the minimum range problem was never raised again.

Stuffy Dowding was probably the only Commander-in-Chief who, during the war years, saw a flight demonstration of two prototype air-interception systems, first the British 200 Mc/s set and then the US 10 centimetre air-interception radar. On his return to Washington, he wrote to General Cheney of the Army Air Corps telling him how well the equipment had performed and urging him to press ahead with manufacture. Within a few weeks of his arrival back in Britain, the

Ministry of Aircraft Production sent the following cable to the British Air Commission in Washington:

> To: British Air Commission, Washington.
> From: Ministry of Aircraft Production
>
> MAP. 6457 9/6. Despatched P.S.3. 9.6.41.
>
> Have been much impressed with the results obtained with NDRC Laboratory 10 cms AI. Should much appreciate immediate despatch of ten such sets to this country by air, with 200 more to follow subject to such modifications as we consider desirable after obtaining experience with ten first sets.

Orders were placed with the Western Electric Company for engineered versions of the MIT set. This was probably the first order for centimetre-wave radars of American origin placed by Great Britain, and a great vindication of Tizard's judgement about America's capacity for rapid production.

Back at the Radiation Laboratory, airborne radar projects multiplied and it was not long before there was a long list of sea-search projects for Naval aircraft, as well as those of the Army Air Corps. These were soon multiplied again to include air-gunnery radars, low-altitude bombing attachments and so on. Perhaps an even more important result was that the Navy became interested in centimetric shipboard radars of every type and these too were added to the project list.

Although the need for an air-interception radar within the US was only of low priority, there was now great pressure on getting a working model over to Britain at an early date. Five more prototypes were put in hand at the Radiation Laboratory. They were updates of the B18 installation, incorporating the many improvements coming out of the laboratory work. One set was intended to replace the B18 installation, one for shipment to Britain, a third to NRC in Ottawa as a working example of the American product and a fourth to the Bell Telephone Laboratory. This last set was to act as a prototype for the first production models from their manufacturing arm, the Western Electric Company, as mentioned above.

It seemed to me that the best way of taking the first set to Britain was to install it in an aircraft in Boston, do all the test flying there and take the finished installation over to England in one unit. If the aircraft could not itself fly the Atlantic, there were now very efficient means of transporting aircraft across the Atlantic by ship. One attractive possibility was to fit a Douglas A20, an aircraft which was already being delivered to the RAF as the Havoc. These were being equipped with AI Mark IV when they arrived in Britain and were giving a good account of themselves as nightfighters. Unfortunately, the delivery schedule of British machines did not fit in well with the timetable, so this did not come about. At this point the ever-willing Canadians came to the rescue and offered a civilian Boeing 247D which they were using for experimental work at the RCAF test station at Rockcliffe.

This was flown to Logan Field and installation of the new set began. At this point, I was extremely lucky in having Dale Corson assigned to take charge of the project. He later figured prominently in many of the Laboratory's activities and then played an important part as design engineer for the low-altitude bombing project, which played a decisive role in interrupting Japanese supply lines in the Pacific. Under his guidance, the installation in the 247D proceeded apace. The performance was improving all the time and we were more than satisfied with a series of flight tests out of Logan Field in Boston. In early June, the 247D was flown to Newark Airport where it was cocooned in preparation for the Atlantic crossing and shipped out of Hoboken as deck cargo. Dale Corson and I flew over to Britain on a B24 of the North Atlantic Ferry on 28 June 1941 and embarked on a series of demonstration flights at Christchurch and FIU.

This was just ten months after the Tizard Mission had left Liverpool in August 1940. The 10 centimetre magnetron had been transported to the USA and been manufactured in large quantities. The Radiation Laboratory had been formed and was in full operation and a wide variety of centimetric radars were being manufactured. In less than one year, one of these sets was delivered to Britain, already installed in its own aircraft. One wonders whether such a timetable would be possible today.

The Boeing 247D in Britain

The Boeing 247D was delivered from the port of arrival to the Fighter Interception Unit at Ford towards the end of June 1941. This was a remarkably fast and smooth operation, considering that Britain had been at war for two whole years. The air-interception gear was soon run up and a series of flight tests and flight demonstrations followed. These were given to three different groups.

(1) The first demonstrations were given to representatives of the centimetre-wave group at TRE. These included Alan Hodgkin and his colleagues in the airborne group, but I cannot remember whether Dee or Lovell flew with the equipment or not. Frankly, the TRE group were lukewarm about the whole thing. They had forgotten, or perhaps did not appreciate, that all the essential transmitting and receiving valves in our early 200 Mc/s radars had come from America and they did not want to believe that the United States was anywhere in the

picture—a curious denial of Tizard's thesis that America had much to offer.

(2) The second group to have a demonstration was the flight crew of FIU itself, of which Jumbo Ashfield, an old friend, was still a prominent member. Their attitude tended to be light-hearted. We have to remember that in the spring of that year they had been through the worst phase of a German night bomber attack and they had beaten it back. The equipment which had done this—the Beaufighter and AI Mark IV—ranked high in their esteem. At that moment, there were virtually no night attacks on Britain and there was no pressing operational demand for anything better. FIU tested the new equipment extensively and wrote a report about it which was full of approval; but I could not detect an urgent demand for production. It is no criticism to say that, if the war had ended right there, I suspect there would have been a long drawn-out discussion in airborne circles about whether to stick with the equipment which had won the last battle or go for a new and revolutionary design on centimetre waves.

(3) We also gave demonstrations to the higher echelon of Fighter Command and the reaction from them could not have been better. They acknowledged the need for continued development and they were very conscious of the fragility of the British supply position. The RAF already depended heavily on supplies of all kinds from America. So, if the Americans thought they could come up with a better air-interception radar, the RAF and MAP were not going to turn a blind eye.

At the technical level, by far the best response at TRE came from H W B Skinner. He had probably done more than anyone else to up-grade British receiver technology. He was intensely interested in the performance of the Radiation Laboratory transmitter and receiver, as compared with the corresponding TRE units. Comparative tests were made and an interesting anomaly emerged. As already pointed out, when the Tizard Mission arrived in America, the US were well ahead of Britain on receiver design but had nothing remotely comparable to the British resonant magnetron. By July of 1941 the position was exactly reversed. The American 10 centimetre pulse transmitter in the Boeing 247D was clearly ahead of its equivalent in Britain, but the British receivers were better. Obviously, each group had been striving to improve the weakest link in the chain and each had made considerable progress. I like to think that both sides profited from the exchange.

I returned to the USA in October and production of the equipment had already started at the Western Electric Company. This was first in prototype form, as the SCR 520, and then in its final form as the SCR

720. A total of many thousands of these were finally produced. Such was the interest in Britain that a Beaufighter was sent out to the Radiation Laboratory flight at Boston and a test installation carried out, first of the SCR 520 set, then the SCR 720. I carried out a series of interception trials in conjunction with the Army Air Corps at Mitchell Field in Long Island, which showed that it had a very satisfactory range of up to 8 miles at 10000 feet. Among other things, these tests showed that interceptions did not have to commence in the stern chase position, but that it was comparatively easy to carry out interceptions from the beam position, that is, when the fighter approached the track of the raiding aircraft from 90 degrees.

Figure 12.4 A Bristol Beaufighter fitted with the Western Electric SCR 720 air-interception radar, seen with a civilian crew at Logan Field in Boston. This equipment went into service with the Royal Air Force as AI Mark X and remained the standard RAF air-interception radar until 1957.

This whole installation was shipped to Britain for further trials, but it turned out that by that time the Mosquito was the favoured night fighter and all subsequent fittings were made in that aircraft. As used by the RAF, the American SCR 720 was renamed AI Mark X and, due to its superior performance against enemy jamming, was finally adopted as the standard RAF air-interception equipment. It remained the standard until 1957. As already described in Chapter 7, these were the radars which played a significant part in intruder operations over Germany and in the battle against the V-1 missiles. Squadron Leader Brandon, in his book *Night Flyer*, which has already been referred to, had this to say about the introduction of AI Mark X:

> . . . after my first couple of trips, I was really delighted with Mark X AI; the more I tried it, the more I realized that it was exactly what we needed from the start: long range, good coverage, reliability, an excellent beacon facility for homings, and it was difficult enough to use properly to make it a constant challenge.

Centimetre-wave Air to Surface Vessel Radar

In parallel with this development, an intensive effort went into the design of a corresponding sea-search system and several varieties were produced. Among the first of these was one which also came from the Western Electric Company—the SCR 717—which followed closely the design of the SCR 720 and had a range–azimuth or B scan. Several additional brands of sea-search equipment were designed with a Plan Position Indicator, or PPI display. Among the first of these was the DMS 1000 which Denis Robinson, a recent arrival from Swanage, engineered in conjunction with the Philco Company. The design was later consolidated into similar sets in the APS and APQ series of the US Navy and Air Corps. These were installed in the B24 and many other aircraft types and were put to great use in the Atlantic and the European theatre. The part played by this type of radar in the concluding stages of the war against the submarines has already been described in Chapter 6. This was the factor which finally ended the Battle of the Atlantic very decidedly in the Allies' favour.

The centimetre-wave ASV radar systems played a similar part in the defeat of the Japanese in the Pacific, but in an entirely different way from the Atlantic battle. In the Pacific, Japan was the nation which was short of indigenous resources and had to maintain a large merchant fleet to supply her home industries and armed forces, which were spread all over the Pacific. This fleet became prey, first to US submarines and then to radar-equipped aircraft. A potent force in this effort were the B24s, equipped with SCR 717, which ranged all over the south-west Pacific and the China Sea. Instead of the Leigh Light, they were equipped with a radar blind-bombing attachment designed at the Radiation Laboratory, called the LAB, or low-altitude bombing attachment, which allowed completely blind attacks on any vessels moving in those seas. Since the Japanese had already made a practice of moving most of their supplies at night, this soon crippled the Japanese supply system. Between them, the submarines and the aircraft practically wiped out the Japanese merchant fleet, and this was an essential preliminary to final victory.

Project 2: Gun-laying Radar—SCR 584

Project 2 of the Radiation Laboratory was the centimetre-wave gun-laying radar for which Cockcroft wrote an outline specification. This was not my field, but in Cockcroft's absence Brigadier Wallace and I handled the early discussions with the Loomis Committee and with the Radiation Laboratory. Cockcroft's specification laid down things like pulselength and pulse power, the desirable maximum range of detection and the range and tracking accuracy. It was quite clear that these figures could be achieved with a modest increase in dish size—from the comparatively small 30 inches used in aircraft applications to an antenna, say, 6 or 8 feet in diameter.

Two decisions which were made quite early in the development played a vital part in the subsequent success of the project. The first of these was the use of conical scan to give precise azimuth and elevation. I have never been sure where this suggestion came from but there were some astronomers among the early recruits at the Radiation Laboratory and it may have come from their acquaintance with optical-tracking devices of a similar kind. I greatly doubt whether it was in the original specification. The second departure from orthodox practice was to abandon manual tracking and go for fully automatic tracking from the outset. As far as I am aware, all Naval and anti-aircraft gunnery sets up to that time had used manual tracking, one operator for azimuth, one for elevation and yet another for range. The advantages of an automatic tracking system were obvious and this was the objective from the start of the project.

There were two other systems under development at that time. These were the Canadian GL3(C) equipment, designed at the National Research Council in Ottawa under Wallace's guidance, and the British GL Mark III, which Cockcroft initiated on his return to Britain; both had manual tracking. The Radiation Laboratory decision to go for automatic tracking from the outset was a far-sighted one and it contributed enormously to the superiority of the set which was ultimately evolved—the SCR 584. Many thousand SCR 584s were built and they first proved themselves in North Africa and at Anzio. A total of 39 sets were landed in Normandy on D-day and they gave a brilliant account of themselves protecting Allied ground forces from enemy air attacks during the subsequent advance through Europe.

In Britain, perhaps they are best remembered for their role in countering the ghastly threat to London and adjacent towns from the

V-1 missiles. Flying as they did at constant speed and constant height, the V-1s were a relatively easy target to the SCR 584. As we have described, aircraft used to attack the missiles as they left their launch sites in France, but they were not always successful. The SCR 584s, coupled with the proximity fuse, wrought a heavy toll on those flying bombs which succeeded in crossing the Channel.

The SCR 584s were deployed around Dover in July 1944 and first came into operation on 19 July. Of the V-1s which reached the coast, the number destroyed by gunfire during the next seven weeks was as given in the table below.

Week	Flying bombs destroyed (%)
1	17
2	24
3	27
4	40
5	55
6	60
7	74

The attacks finally ceased when the Allied Forces overran the launch sites during their drive through France.

Project 3: Long-range Navigation—LORAN

I have already mentioned Tizard's insistence that the North American continent was a much better place than war-torn Europe to develop and test a long-range navigation system. Loomis fully appreciated this and practically overnight—on the basis of a description I had given him of British GEE—came up with the suggestion of doing a similar thing on a longer wavelength which allowed over-the-horizon operation. Responsibility for this project fell on Melville Eastham—the genial owner and General Manager of the General Radio Corporation of Cambridge, Massachusetts—the maker of some of the best radio measuring equipment ever made in the 1930s and 1940s. He was older and considerably more experienced than the average Radiation Laboratory staff member and brought a maturity to Project 3 which was unusual in that very young organisation.

The initial proposal was to erect three ground stations about 1000 miles apart and, as early as November 1940, sites were being considered in Nova Scotia, at Blue Hills just outside Boston and in Maryland. There were some changes of plan about the most useful wavelength, which were mainly dictated by available gaps in the frequency spectrum, but they ultimately settled on 2 Mc/s—a frequency at which the transmission characteristics were well known and predictable. The project was first known as LRN, standing for long-range navigation, but it was soon changed to LORAN. After the first year or so, Jack Pierce of Harvard University took over leadership of the group. He had long experience of ionospheric work and had an important impact on the remainder of the development.

The first operational system came into being in 1942 and covered the North Atlantic. By 1943, the whole of Europe was covered, east to the Danube and south to the North African coast. By the end of the war, the coverage had been extended over the whole of the Pacific, including the all-important supply routes into Burma, India and China—a total area greater than one third of the surface of the globe.

Achievements of the Tizard Mission in the USA

As we have described, there were many subjects on which the Mission disclosed recent British advances and received in return a wealth of information from the US—on jet engines, rockets, predictors for Naval and anti-aircraft gunnery, bomb sights, proximity fuses and many other subjects. There is a great need for an all-embracing account of what the Mission accomplished in all these fields. Such a story might be written from two points of view—what was given and what was received by Great Britain, and what was given and received by the United States. There must be a vast store of documents from which that story could be assembled. In what follows I will attempt to summarise the radar exchanges which took place.

(1) The Tizard Mission found that the USA had already made substantial progress with many forms of radar; for example the CXAM at the Naval Research Laboratory, which was virtually the equivalent of British CHL and operated on the same frequency. Similarly the SCR 268 and the SCR 270 at Fort Monmouth. But the US sets had only been built in small numbers and had not seen much operational use. Radar scientists at both the NRL and Fort Monmouth establishments had fought an up hill battle for recognition of the importance of their work and the need for it to go into service. The Tizard Mission changed this in a dramatic way. It convinced the US Chiefs of Staff that radar was a supremely important weapon of war, one which was

already playing a vital part in military operations in Europe. The very fact that the Battle of Britain was being won while these discussions were taking place was conclusive proof of the point made by the Mission. Production orders for US radars and orders for US equivalents of British radar equipment followed in great profusion.

(2) Probably the greatest single achievement of the Tizard Mission was to introduce the resonant magnetron to the USA. It was subsequently produced by the million and was incorporated in hundreds of thousands of radar sets of every description. In his book *Scientists Against Time* (1946), J P Baxter says of the magnetron 'When members of the Tizard Mission brought one to America in 1940, they carried the most valuable cargo ever brought to our shores.'

(3) The introduction of the magnetron led to the formation of the Radiation Laboratory at MIT. From small beginnings in October 1940, the staff of the Radiation Laboratory grew to a total of nearly 4000 people by the end of the war. It designed radar equipment for the US Army, the Navy and the Air Corps which was produced in hundreds of thousands, to a value of billions of dollars. These were used in every theatre of war and brought about a complete transformation in the concept of military radar, from the comparatively long wavelengths used in the late 1930s to the centimetric radars of the present day.

The Radiation Laboratory was a new and unique concept in American science. It was a government-funded enterprise, placed at a university and staffed by the most highly qualified scientists who could be found in the whole of the United States. It was formed in close consultation with industry and remained hand-in-hand with them for the duration of the war. It could not fail to be an outstanding success and it became the model on which many other such laboratories were based. It owes a tremendous debt to the genius of Alfred Loomis, to Vannevar Bush and Karl Compton and to Lee DuBridge, its Director; but the lead given by Tizard must not be forgotten.

(4) Last but not least, the Tizard Mission introduced airborne radar to the United States. This was a field in which Britain was undoubtedly ahead of the rest of the world. The USA was quick to follow this up and the Germans never caught up with this branch of technology.

The contribution of the Tizard Mission can be looked at in a slightly different way. It established a rapport between the USA and Britain which enabled them to talk to each other in a relaxed and understanding way about common military problems and about the problems likely to be encountered in the future. This process was formalised by the formation of the British Commonwealth Scientific Office in

Washington as the natural successor to the Tizard Mission. It was followed by the corresponding NDRC (later OSRD) Office in London in 1941. These set a seal to the accord between the two countries and between the two governments. It was Tizard's vision which showed the way and demonstrated that this kind of collaboration was possible.

13

Postscript

By the end of 1943, the war situation was much more hopeful than it was in the dark days of 1940. There were clear signs that the tide of battle was turning in favour of the Allies, both in Europe and in the Pacific. In Europe, German forces were in retreat in the Mediterranean and under tremendous pressure from Russia on the Eastern Front. Italy had surrendered and preparations for the invasion of Europe were under way. There were indications that the Battle of the Atlantic—covering the vital supply route from North America to Britain—was on the way to being won.

In radar itself there had been a distinct change in emphasis; it was no longer regarded as a purely defensive weapon and serious consideration was given to how it could be used in offensive roles. By mid-1943 most of the radar devices likely to be used, either in the defensive or the offensive mode, had already been thought of or were being worked on; it was apparent that putting these into practical use was much more important then the development of entirely new devices. Only a few years before, practically any proposal genuinely capable of contributing to the war effort was seized on by the Armed Forces and research initiated almost without question. Now a rather more pragmatic philosophy came into play. Could any new radar proposal, however attractive technically, be brought to completion and put into service soon enough to contribute to the war effort? In this context, it became clear that my task in the USA was drawing to a close and I began to give serious thought to the direction in which my next efforts should be aimed. There was little more that I could do within the United States—the radar tide was already in full flood. I took advice from several sources and an important one of these was Tizard himself. He always kept his ear close to the ground and came up with a typical piece of good sense. He said 'They seem to need help in Australia. Go there, my man.'

It happened that back in the spring of 1939, well before British radar was disclosed to the USA, the Committee for Imperial Defence decided that Commonwealth countries should be informed of radar. Australia, Canada, South Africa and New Zealand were told that a vital new development in defence science had taken place and that it was in their interest to send people over to Britain to learn about this at first hand. They responded immediately and senior scientists came to Bawdsey to learn the secrets of radar in some detail. Each of these countries was encouraged to embark on research for themselves and to consider which brands of radar to put into use for their own protection. As a result, experimental work started in all four countries. In Canada, this was centred at the National Research Council at Ottawa; in Australia, the Radiophysics Laboratory was formed in Sydney to conduct the work.

Australia had a relatively long history in radio research. It began in the early 1920s when intensive efforts were made to establish a direct radio link with Great Britain. Interestingly enough, this was entirely an amateur development; it was centred at the Mill Hill School in London and corresponding groups in Australia; they attempted to communicate with each other on what were then called the ultra-short wavelengths, thought to be of no commercial value. Their efforts were successful and the provision of a proper service was taken up by the respective governments and by commercial interests. A thriving radio industry grew up in Australia; some companies were indigenous, while others were subsidiaries of well known British, European and American companies. The structure of the industry was modelled on that of Great Britain, with a strong Post Office interest and connections to the Australian Army, Navy and Air Force. A Radio Research Board was also formed, based on the British model; using grants from governmental sources, the Board became active in stimulating research in universities and other institutions.

When the need for radar research arose, Australia was therefore in a good position to forge ahead. This was done by setting up a special laboratory—the Radiophysics Laboratory—in Sydney under the wing of the Council for Scientific and Industrial Research (CSIR). This was itself a government body, modelled on the Department of Scientific and Industrial Research in Britain. There were also close links with Britain at the university level. Australia's best young scientists were traditionally given studentships to Cambridge and other universities. Some of Appleton's best PhD students came from the Dominions, and names like Barnett, from New Zealand, and Builder and Pulley, from Australia, appear as co-authors in many of Appleton's basic papers about the ionosphere. One of the lecturers in physics during my time at King's College was F W G White from New Zealand and in 1943 he was the head of the Radiophysics Laboratory in Australia. He was destined to move up through the hierarchy of CSIR and later became Chairman of the organisation. These were all close friends of mine when I was a student at King's College. I had worked with them, dallied with them in London pubs and was present at more than one wedding when they finally decided to take the plunge and marry their English girlfriends.

I had never previously been south of the equator, but when Tizard suggested I should move from the USA to Australia I already had many friends there and readily agreed to go. I left the United States a few days after Christmas 1943 in the bomb bay of a US Air Force B24, flying out of Hamilton Field in San Francisco. The trip was marginally better than those I had made back to Britain on the North Atlantic Ferry in the previous few years. Those flights were made at 14 000 feet, lying on mail bags and miscellaneous stores in the bomb bay and travelling over a freezing cold Atlantic clearly visible through gaps in the bomb doors. For the long Pacific crossing there were real seats and although the sea could still be seen through the floor, it was comforting to know that its temperature was about 20 degrees Celsius and just right for a nice swim. But it was still a wild ride and I got the distinct impression they used cowboys as pilots. After stops at exotic places like Hawaii, Canton Island and Noumea, I arrived in Sydney on 1 January 1944.

The Radiophysics Laboratory was housed in a building specially built for the purpose in the grounds of Sydney University. It was already a flourishing concern with a staff of two hundred people; it had an excellent mix of highly qualified scientists and engineers and the permanent building included first-rate workshop facilities. For its size, it was one of the best equipped laboratories in which I had worked up to that time. The laboratory had good connections with industry and in spite of the enormous distances and the difficulties of wartime travel, it had kept closely in touch with radar research in Britain and America. Members of the staff made frequent visits to TRE and MIT and were familiar with the latest technology.

The Laboratory had responsibilities to all three Services, the Army, Navy and Air Force and to the US Forces in the south-west Pacific, which had their headquarters in Brisbane under the command of General MacArthur. A noteworthy product of the Laboratory was a lightweight air-warning radar, called the LW/AW, which was specially designed for island hopping and could be carried in by the troops on landing barges or in transport aircraft. Many hundreds were produced and were used with great effect by both the Australian Forces and the US Army. This set played a significant part in the success of the island-hopping campaign against the Japanese. Apart from special products of this nature, the group had also embarked on a small production run of ASV Mark II and were already producing copies of the resonant magnetron for radars which operated over a range of wavelengths from 3 centimetres to 30 centimetres. Nominally, my own part in all this was to be in charge of research but, as we have seen, there was little need for new initiatives at that stage in the war and my task was mostly to provide help with the practical application of the radar sets I already knew so well.

In the Pacific, the Japanese forces had advanced northward as far as the Aleutians; in a southerly direction they had come within striking distance of Australia before they were finally stopped on the Kokoda trail in New Guinea. After grievous losses in the early part of the war, the US Navy was beginning to take control in mid-Pacific and the fight back was about to begin on land and at sea. As we have seen, there were profound differences between the Pacific battle and the one in Europe. The Japanese were short of natural resources and almost completely dependent on raw materials imported from all over the Far East. In the same way, Japan had to supply her armed forces over extremely long lines of communication, extending 7000 or 8000 miles from the Aleutians to New Guinea. This required a huge merchant fleet which had been painstakingly built up before and during the war. This fleet became prey first to US submarines and then to radar-equipped aircraft, both land and carrier based. In the waters of the South Pacific, aircraft adopted a policy of hitting anything that moved and, between them, the submarines and the aircraft practically wiped out the Japanese merchant fleet. Of 300 000 Japanese troops which landed in New Guinea, only 100 000 survived. There had been some fierce fighting, but their losses resulted just as much from starvation and disease as from casualties in battle. The story was similar elsewhere in the Pacific and the subsequent fight became one not of frontal assault, but of island hopping. Whole battalions of Japanese were isolated in remote parts of the Pacific and their supplies completely cut off.

As the process continued up through the Philippines and to within reach of the main islands of Japan, it was only a matter of time before the knock-out blow was administered. As we now know, with the advent of the atomic bomb the final event took an entirely different form. It brought about a sudden end to the war, with a terrible loss of Japanese lives which will always be regretted. But it is equally true to say that if the bomb had not materialised at least as many Allied lives would have been lost before the war was finally won.

Impact of Radar on Post-war Activities

At the end of the war, in winning and losing countries alike, it was time for a massive rethink about the direction which future research should take. There was clearly a need for a general resumption of civilian research, not only along branches which had been neglected during the war, but along entirely new lines which had been thrown up by war work. We knew of any number of these, which the urgency of wartime work had prevented us from following up.

Radar introduced some new and revolutionary concepts into warfare, and in doing so gave birth to entirely new branches of technology; it was inevitable that these would continue after the war and have a widespread impact on many aspects of civilian life. Some of the changes came about by a direct transfer of radar technology to everyday activities like the navigation of ships and the guidance of civil aircraft. As always, there were also contrary forces; there is the famous story about a Naval Commander in the British Pacific Fleet, lying in Sydney Harbour at the end of the war. As the youngest and keenest radar officers were being paid off, he looked up at the tangle of radar antennae at the masthead and said: 'Now we can get rid of all that garbage and get back to being a real Navy.' No doubt there was also a sprinkling of Army officers who looked forward to getting their horses back. But the prevailing mood was different; anyone who had been in any way involved in the new technology saw innumerable ways of feeding it back into the everyday activities of mankind, and even into fundamental science itself. Whether they were chemists, physicists, biologists or physiologists, when they went back to normal life from wartime radar they carried with them a wealth of new knowledge, new instruments and new techniques which had not previously been available.

Radar and Marine Navigation

At the end of the war, the LORAN long-range navigation system covered over a third of the earth's surface, including the busiest parts of the Atlantic and practically the whole of the Pacific. It was taken up at once by the merchant fleets, simply by using military style equipment without modification of any kind. This was so successful that there were many competing systems and a healthy market arose for new and improved methods of long-range navigation.

Another marine application was in the use of military PPI- type radar systems for navigation of ships in coastal waters and in inland waterways. The ability of such radars to detect buoys, other ships and obstructions in closed waters, particularly under conditions of low visibility, introduced a new dimension into maritime navigation. These devices quickly spread through the merchant fleets of the world and today it is unusual to find even a cabin cruiser or the humblest fishing vessel without its radar antenna. The worldwide market was immense and was catered for by a wide range of manufacturers in almost every advanced country.

Radar Aids to Civil Aviation

In the same way, radar was adopted to assist in the navigation of civil aircraft. Once again the LORAN- type system was much used by long range aircraft on the main ocean crossings, supplemented by PPI-type airborne radar which had a two-fold purpose. It provided navigational assistance by identifying the terrain ahead of the aircraft at night or under conditions of low visibility; it was especially useful in regions where there were distinctive land–water boundaries, as in the south-west Pacific. It provided an added bonus by identifying heavy rain and storm areas which might prove hazardous and therefore to be avoided. Today, there is hardly a passenger-carrying aircraft anywhere in the world which does not have such a radar in its nose.

Another group of navigational aids which evolved after the war was the various types of track guidance system, aimed at guiding aircraft in the critical last hundred miles or so of their run into an airport. A feature of wartime radar was the precision with which such tracks could be defined; accuracies of a few hundred yards at a distance of a hundred miles were not unusual, and in post-war applications this was soon improved upon. It meant that an aircraft could place itself on an approach or landing path with great accuracy, and many systems of this kind came into use.

Yet another instrument depended on the ability of radar to measure distance to an extraordinary degree of accuracy. A very simple piece of equipment, specially developed for airline use, was the so-called Distance Measuring Equipment or DME. This provided pilots with a direct measurement, in miles, of the distance to the terminal airport, or the distance from an airport they had just left. My own group at the Radiophysics Laboratory in Australia set about producing such an aid. The Australian version of DME was first used on the country's air routes in 1950 and in 1954 it became mandatory equipment on all passenger-carrying aircraft in Australia. It was almost 10 years before similar equipment appeared on the air routes of the northern hemisphere. The same equipment has now been updated, miniaturised and transistorised; it is still in use on Australian air routes and is scheduled to continue in service until at least the 1990s.

Along with these aids, ground-based radars were developed for the control of aircraft along the air routes or in the airport control area. These were strictly analogous to wartime air-warning radars, giving the position of all aircraft within the jurisdiction of an airport. These sets have proved of inestimable value in the sequencing and the control of air traffic, particularly under bad weather conditions; they have been especially helpful in the regions of high traffic density around capital cities. There is no doubt that, even in clear weather, present day landing rates could never have been achieved without the widespread use of such systems.

Radar Aids to Surveying

The ability of radar to measure distance to a high degree of accuracy was exploited during the war in devices like the OBOE blind bombing system. This was a method by which a Pathfinder aircraft over the Ruhr was positioned accurately by measuring its distance precisely from two ground stations in Britain. An interesting fact soon emerged. Conventional ground survey methods had reached a high degree of sophistication well before the war and these had been used to link the British survey to that of France, and that of France to that of Germany. It turned out that the accuracy with which this had been done was much poorer than could be measured by the OBOE system. The maps had, therefore, to be corrected, after which an aircraft could then be placed with an accuracy of 20 or 30 yards over particular factories in the Ruhr.

The existence of such mapping errors was not forgotten in the immediate post-war period. Australia, for example, is a continent with an area of which over half is inaccessible and difficult to traverse, let alone survey with any precision. It was known that even along the relatively well travelled Adelaide to Darwin route there were occasional errors, not of yards, but of many miles in surveys starting from the north as compared with those coming from the south. Up to that time, the traditional methods of survey were almost entirely based on triangulation, that is, building up a series of triangles from a single measured baseline. The angles of each triangle were measured to a high degree of precision but only occasional checks of distance were possible when the terrain allowed it. In remote or inhospitable country, such checks simply could not be carried out and the accuracy suffered accordingly. Radar provided an entirely new dimension in that the triangulation could be established by the accurate measurement of distance alone, made in sections of 200 miles at a time. It was not even necessary to traverse the intervening ground; an aircraft could fly over the terrain, position itself from two radar beacons on the ground and simultaneously record its position by vertical photography.

These methods were of inestimable value in remote parts of countries like Australia and Africa which might contain minerals and other material needed by a resource-hungry world. It is probably true to say that radar survey methods were not much used in more developed countries which already had reasonably accurate surveys, but they were enormously important in less developed parts of the world.

Radar in Meteorology

We have already mentioned the ability of a radar to spot areas in which rain has formed; this is especially true of sets operating on wavelengths around three to five centimetres. Ground radars of this type are now in routine use by meteorological services around the world to give warning of approaching rainstorms, and they have become an accepted part of any television weather report.

In the same way, airborne radars of this type proved to be a great help to aircraft by allowing them to recognise and avoid potentially hazardous storm areas. However, the utility of this type of radar went far beyond this in that it could also be used to assist research into cloud and rain physics. It is strange to note that even in the 1940s the exact mechanisms which lead to the formation of rain were not well understood. Radar-equipped research aircraft played a significant part in sorting out this particular problem. They would observe the radar echoes inside a cloud as rain developed and then fly through measuring the water content, the ice and water droplet count, upward air currents and so forth. There was great activity in this field in many countries and within 10 or 15 years a much better understanding was gained of how and why rain forms.

Most of the simpler aspects of rain formation are now well understood and research has tapered off. However, in the case of very damaging storms, like hurricanes or the extreme hailstorms which tend to occur in certain geographical regions, much work remains to be done. This is carried out predominantly by radar-equipped aircraft. The US Weather Bureau maintains several Lockheed P3 aircraft which are specially strengthened to allow them to fly directly through the eye of hurricanes off the Florida and California coasts. They carry very advanced radars and sophisticated instrumentation and have saved many lives by pin-pointing just where and when such storms are likely to hit inhabited regions.

Electronic Computers

A feature of the last 30 years, in industry, in commerce and even in the home, has been the proliferation of electronic computing machines. This is a development which can be traced back not so much to radar itself as to the enormous advances in circuit design which were an integral part of radar research during the war. Ever since the abacus of Babylonian times, man has dreamed of constructing machines which would assist in the intricate calculations required for trade and commerce. Many attempts were made to build such machines, but with indifferent success until the nineteenth century. Then in the 1850s, Babbage and his talented protégé, Lady Ada Lovelace, were probably the first to see exactly how such a machine could be built. Babbage made plans for an incredibly complex machine of this type and spent a lifetime attempting to build it. Like many innovators, before and since, he ran into much opposition and was often ridiculed for his attempts. It is now clear that he was on the right track and was only defeated by the sheer complexity of the gears and mechanical devices which had to be built, and the precision with which they had to be constructed.

Further attempts at solving the problem, still by mechanical or semi-mechanical means, continued into the 1930s with varying degrees of success. During the war, some elements of electronic computing had been built into gunnery predictors and bombsights, but not until the war ended was it realised that a whole range of computing operations could be performed electronically. One of the most intriguing things electronics had to offer was the ability to perform at speeds well beyond the capacity of mechanical devices. Microsecond pulses were commonplace at the end of the war and this alone held the prospect of being able to count pulses at a rate of a million a second. Today, speeds ten times faster are available in home computers and speeds fifty or a hundred times greater are in regular use in more advanced machines.

By the end of the war, a whole range of pulse-forming networks, pulse-counting circuits and memory circuits existed in the form needed for an electronic computer. The race began and many centres became active in the field, notably Princeton and MIT in the USA and Cambridge and Manchester Universities in Britain. Today, a whole spectrum of calculating machines exists, ranging from giant instruments dedicated to scientific research, to intermediate machines for industrial and business functions, down to the minicomputers found in many homes. This is truly a case in which wartime radar has paid back its debt to pure science many times over.

The speed with which these developments took place is simply astounding. Machines which filled whole rooms and cost hundreds of thousands of dollars in the 1950s were scaled down to the size of desk-top machines a few years later; today, the same capacity is available in hand-held calculators costing a few hundred dollars. An intriguing fact is that the secrets of operating even the most arcane machines are not confined to adults with special training. Children of quite tender age seem to be completely at home with these machines and sometimes put their elders to shame with their expertise. In my own childhood, a telephone was a rare thing in the average home, and the thought of talking to someone hundreds of miles away or in another country was sheer fantasy. One wonders what the impact on society will be when today's computer prodigies reach adulthood.

Radio Astronomy

Yet another example of radar technology being fed back into fundamental science is found in the post-war resurgence of studies in radio astronomy. This science began with an observation by Jansky in 1933 at the Bell Telephone Laboratory at Holmdel, significantly enough the very same laboratory at which Englund, Crawford and Mumford made their observation of reflected signals from aircraft. By studying the noise coming into a receiver used to monitor transatlantic signals, Jansky showed that while much of the noise was of locally generated electrical origin or came from natural phenomena like thunderstorms, another part must have come in from outer space. He called this cosmic noise. Little notice was taken of this observation at the time and it lay almost fallow until the end of the war.

As radar men, we were interested in similar questions and were naturally concerned about cutting the noise in our receivers to a minimum. Some of this noise was generated in the front end of our receivers, but we knew about Jansky's work and from the very first days at Orfordness and Bawdsey we recognised that when our antennae pointed at a target in the sky—in this case an aircraft—we would also receive a noise component from outside the atmosphere. We called it Jansky noise, by analogy with the Johnson noise and the Schott noise generated in the receiver itself.

During the war, there was sporadic mention of radio noise coming from outside our own atmosphere. J S Hey, in England, observed that an unusual amount of radio noise seemed to be coming from the sun, but only at times of sunspot activity. Little notice was taken at the time and it did not become generally known until after the war. Then Southworth, a colleague of Jansky's at the Holmdel Laboratory, measured the surface temperature of the sun with a 10 centimetre receiver and obtained a value sufficiently close to the known temperature of 6000 kelvin to excite interest. Lastly, Dicke, using a supersensitive receiver on 1.25 centimetres which he built at the Radiation Laboratory, measured the surface temperature of the moon—a mere 250 kelvin†. None of this was published until the war had finished, but as a resident at MIT and a frequent visitor to Holmdel I was familiar with them and wondered what might be done about it after the war.

It is little wonder that it was radar people who pioneered the rebirth of radio astronomy as soon as there were opportunities to do so. Among the most active of these was a group which grew up under the guidance of Lovell at Manchester University, another under Ryle at Cambridge University and still another, under the benign guidance of Joe Pawsey, in my own group at the Division of Radiophysics. All of these people were directly connected with wartime radar. The full

† $250\ K \simeq -23°C\ (0°C \simeq 273\ K)$.

Figure 13.1 The six foot diameter antenna of the SCR 584 gun-laying radar, as designed in 1940.

story of how radio astronomy developed and the course that it took is far beyond the scope of this book, but a first-hand account of the early post-war years will be found in a book entitled *The Early Years of Radio Astronomy,* edited by Woody Sullivan (1984).

Most of the early studies in radio astronomy were done with ex-radar equipment, crudely adapted to observe signals from the sun or outer space. Gradually, many new instrumental concepts evolved—a wide variety of interferometers, cross antennae and synthesis telescopes, all of which made significant contributions to the new science. In the northern hemisphere, giant steerable telescopes were built, the largest of which at that time was Lovell's 250 foot diameter instrument at Jodrell Bank. In the southern hemisphere, I was involved in building a counterpart in the form of the 210 foot radio telescope at Parkes in New South Wales. Construction of this instrument began in the 1950s and was completed in 1961. The similarity between the design of this structure and that of a wartime radar antenna is shown in figures 13.1 and 13.2, where a photograph of the 6 foot dish of an SCR 584 gun-laying radar may be compared with that of the 210 foot diameter Parkes telescope. The SCR 584

Figure 13.2 The 210 foot diameter dish of the radio telescope at Parkes in New South Wales, photographed in 1961.

antenna was servo-driven and designed to track an aircraft across the sky with a precision of a few seconds of arc. The Parkes telescope was also servo-controlled and, when built, was one of the largest devices ever driven in this way. It was capable of tracking celestial objects across the sky to an accuracy of fifteen seconds of arc—a precision which was unattainable only a few years earlier. It had other unique features, including a dish structure which was self-adjusting to compensate for the effects of distortion due to gravity. When first built, the surface accuracy was already good enough for it to operate at a wavelength of 10 centimetres. In the next few years it was upgraded so that a substantial fraction of the central surface could operate at millimetre wavelengths. When we first put 30 inch diameter dishes in the noses of Beaufighters and Mosquitoes, we had doubts whether sufficient accuracy could be maintained for them to work at a wavelength of 10 centimetres. Much higher accuracy is now possible as a matter of routine in dishes hundreds of feet in diameter.

Today, a wide variety of radio telescopes exists for a whole range of studies of the Sun, the planets and objects in the very depths of outer space. These instruments have many discoveries to their credit, like the quasars, the most distant objects in the universe, and the enigmatic pulsars, which behave like gigantic rotating beacons in outer space. They have been sending out signals almost since the universe began, but no one was aware of their existence until radio astronomers discovered them. Radio telescopes have now taken their place alongside the giant instruments used in optical astronomy and have become an integral part of the way in which the science is conducted. Not only have they given rise to an entirely new branch of astonomy, but they have revolutionised the whole concept of how our universe is built and how it operates.

This story began when the vast storehouse of radio knowledge which had accumulated in the years between 1900 and the 1930s was poured into solving the problem of air defence. From that beginning, it spread into all aspects of military science and during the next decade there was an explosive growth of new devices inspired by wartime needs. Along with this was a striking growth in technology, all of which has now been fed back into basic science and into many branches of commercial and industrial activity. Interestingly enough, this is not a new phenomenon but a process which has probably been going on since man first went to war. We know that Archimedes helped the military with his fireballs and mirrors, and that Galileo tried to improve gunnery by investigating the trajectory of the cannon ball. Rumford provided an excellent example of the process working in the reverse direction when, as a result of careful measurement of the amount of heat generated while boring the barrel of a cannon, he formulated an important law of thermodynamics and fed it straight back into fundamental science. This process has been repeated once again in the twentieth century and, although the latest repayment into fundamental science and into the everyday activities of mankind has now been going on for 40 years, there is no sign of it letting up. I have enough faith in the benefits which scientific endeavour can bring to mankind to think that, although there may be occasional set-backs and hitches along the way, the process will continue for the rest of time.

Appendix 1

Total Production Figures for 200 Mc/s AI and ASV

AI Production

It has been difficult to establish the correct British production figures for 200 Mc/s AI and ASV equipment during the war, but the number ordered each year up to 1941 are as given below:

1939	AI Mark I handmade. Metro-Vickers and Pye Radio	30
1939/40	AI Mark II. E K Cole and Pye Radio	300
1940	AI Mark III and IV. E K Cole, Pye Radio and EMI	3000
1941	AI Mark IV. E K Cole, Pye Radio and EMI	3000

After 1939, the dates for completion of these deliveries would have been eight or ten months later than the dates given above.

ASV Production

The corresponding figures for 200 Mc/s ASV in Britain for the first three years of the war were:

1939	ASV Mark I. E K Cole and Pye Radio	300
1940	ASV Mark II. E K Cole and Pye Radio	3000
1941	ASV Mark II. E K Cole and Pye Radio	3000

An unknown number of these were diverted to Naval and Army use. These were converted to Naval sets Type 286 and the Army searchlight director set Type SLC respectively.

The ASV Mark II was probably the most widely produced radar of the whole war. The known production figures outside Great Britain are:

USA	Philco Corporation	7000
Canada	Research Enterprises	10000
Australia	PMG Research Laboratory	1300

Appendix 2

A copy of the first page of a memorandum to Watson Watt from E G Bowen at Orfordness, dated 9 October 1935, together with a transcript of the text.

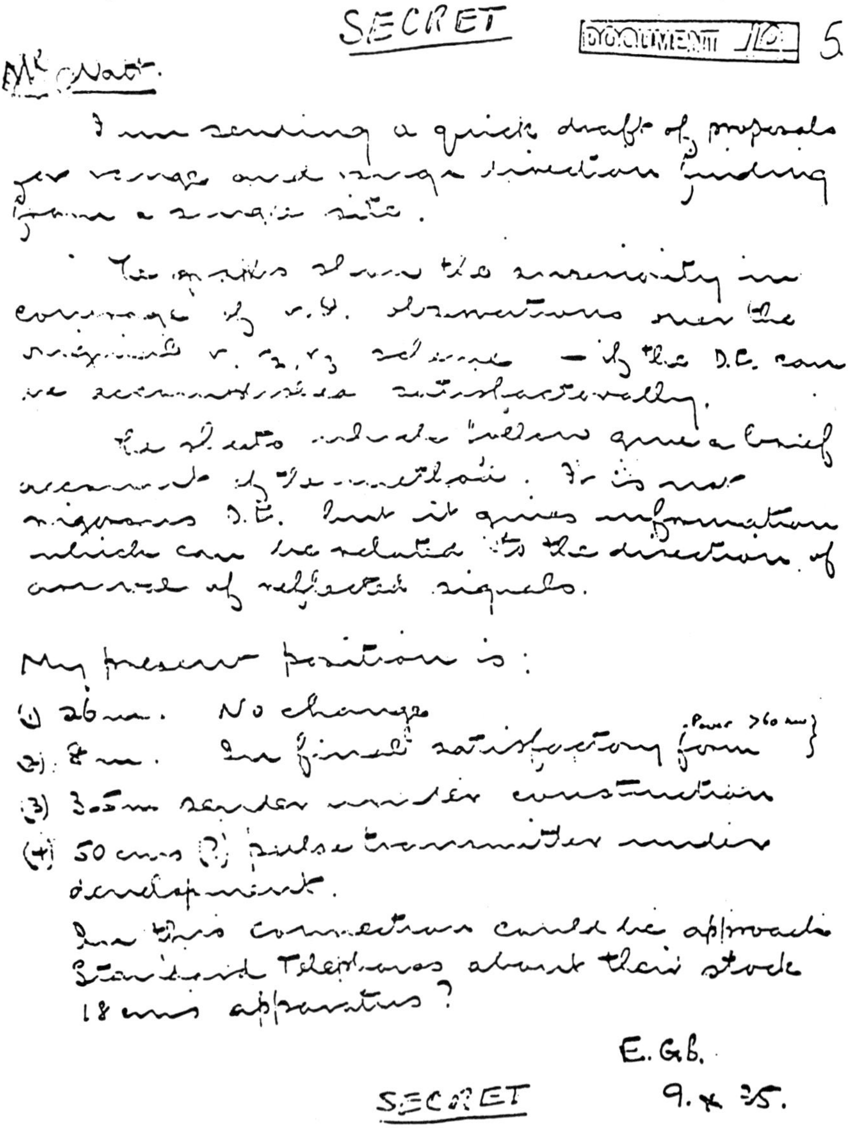

SECRET   DOCUMENT 10   5

Mr Watt.

I am sending a quick draft of proposals for range and single direction finding from a single site.

The graphs show the superiority in coverage of r.f. formations over the original r, r_2, r_3 scheme — if the D.F. can be accommodated satisfactorily.

The sheets included herewith give a brief account of the method. It is not rigorous D.F. but it gives information which can be related to the direction of arrival of reflected signals.

My present position is:

(1) 26m. No change

(2) 7m. In final satisfactory form {Power >60 kw}

(3) 3.5m sender under construction

(4) 50 cms (?) pulse transmitter under development.

In this connection could we approach Standard Telephones about their stock 18 cms apparatus?

E.G.B.

SECRET   9.x.35.

SECRET

Mr Watt.

I am sending you a quick draft of proposals for range and rough direction finding from a single site.

The maps show the superiority in coverage of r, θ observations over the original r_1, r_2, r_3 scheme—if the D. F. can be accomplished satisfactorally [sic].

The sheets which follow give a brief account of the method. It is not rigorous D. F. but it gives information which can be related to the direction of arrival of reflected signals.

My present position is:

(1) 26 m. No change.

(2) 8 m. In final satisfactory form. (Power > 60 KW).

(3) 3.5 m sender under construction.

(4) 50 cms (?) pulse transmitter under development.
In this connection, could we approach Standard Telephone about their stock 18 cms apparatus?

E. G. B.
9. X. 35.

SECRET

Appendix 3

Air Ministry Centimetric Airborne Radar Contract with GEC and EMI

List of Meetings (1940)

Meeting Number	1	16 January
	2	15 February
	3	14, 15 March
	4	9 April
	5	26 April
	6	8 May
	7	22 May
	8	14, 15 June
	9	8, 9 July
	10	19 July
	11	5 August

Complete minutes of these meetings were kept by the GEC Research Laboratories at Wembley, who provided the secretariat. The only minutes which could be found in 1985 were those for 5 August, in AVIA 15/635 and 636 at the Public Records Office, Kew.
Subsequent meetings, commencing 30 August, were held in a different form with Tommy Walmsley of MAP as Chairman and P I Dee as the TRE representative. Details may be found in the relevant Air Ministry files under AVIA 15.

Appendix 4

A copy of an aide-mémoire from Lord Lothian to President Roosevelt about the Tizard Mission, dated 8 July 1940.

Mr. President

Aide-memoire.

The British Government have informed me that they would greatly appreciate an immediate and general interchange of secret technical information with the United States, particularly in the ultra short wave radio field.

It is not the wish of His Majesty's Government to make this proposal the subject of a bargain of any description. Rather do they wish, in order to show their readiness for the fullest cooperation, to be perfectly open with you and to give you full details of any equipment or devices in which you are interested without in any way pressing you beforehand to give specific undertakings on your side, although of course they would hope you would reciprocate by discussing certain secret information of a technical nature which they are anxious to have urgently.

I presume that, if you approve in principle of this interchange of information, you would wish to discuss it further with the War and Navy Departments before giving a decision, and, should you so wish, I would be glad to place my Air Attache and the scientific assistant to the Air Attache at the disposal of the staff of the C.G.S. (General Marshall) and the C.N.O. (Admiral Stark) with a view to their discussing what technical matters might be of interest to these Services.

As to subsequent procedure, should you approve the exchange of information, it has been suggested by my Government that, in order to avoid any risk of the information reaching our enemy, a small secret British mission consisting of two or three service officers and civilian scientists should be despatched immediately to this country to enter into discussions with Army and Navy experts. This mission should, I suggest, bring with them full details of all new technical developments, especially in the radio field, which have been successfully used or experimented with during the last nine months. These might include our method of detecting the approach of enemy aircraft at considerable distances, which has proved so successful; the use of short waves to enable our own aircraft to identify enemy aircraft, and the application of such short waves to anti-aircraft gunnery for firing at aircraft which are concealed by clouds or darkness. We for our part are probably more anxious to be permitted to employ the full resources of the radio industry in this country with a view to obtaining the greatest power possible for the emission of ultra short waves than anything else.

LOTHIAN

8th July, 1940.

Appendix 5

Meetings of NDRC Microwave Committee subsequent to 1 October 1940 as derived from the personal diary of E G Bowen.

14 October 1940—21 East 79th Street, New York City
Present: Alfred Loomis (Chairman)
Professor Ernest Lawrence
Professor Ed Bowles
Mervin Kelly, Bell Laboratories
Ralph Beal, RCA
Hugh Willis, Sperry

Due to the short notice of the meeting, GE and Westinghouse representatives were unavoidably absent.

Notes

Arrangements were made for allocation of contracts for component parts of 10 centimetre air-interception equipment for completion in 30 days as follows:

1 Magnetron—Bell Laboratories
2 Magnet—GE
3 Pulse modulator—RCA
4 Paraboloid and Scanning Gear—Sperry
5 Primary Power Supply—RCA
6 10 Centimetre Receiver—Bell Laboratories, RCA and Sperry
7 Indicator—Bell Laboratories, RCA and Sperry
8 Delay Screen Cathode-Ray Tube—RCA

Discussed organisation of Laboratory Management Committee and personnel to be recruited by Ernest Lawrence.

18 October 1940—Carnegie Institution, Washington DC
Present: Alfred Loomis (Chairman)
Vannevar Bush, Chairman NDRC
Karl Compton, President MIT
Professor Ed Bowles, MIT
Dr Coolidge, Director of Research GE
Dr Jolliffe, Director of Research RCA
Ralph Beal, RCA

Hugh Willis, Sperry
George Metcalf, GE
J A Hutcheson, Westinghouse
Commanders Briscoe and Tucker, US Navy
Majors Mitchell and Smith,
US Signal Corps

Notes

Settled contract arrangements for 10 centimetre airborne radar:

1 Magnetron—Bell Laboratories, 5 in 30 days
2 Magnet—GE, probably 1500 gauss at 2.5 pounds weight
3 Paraboloid and Scanning Gear—Sperry in 30 days
4 Primary Power Supply—Bendix
5 Pulse Modulator— RCA, 1 in 90 days
Westinghouse, 1 in 30 days
GE, estimate later
6 Receiver— Bell Laboratories, 5 crystal receivers in 30 days
RCA, 2 receivers in 90 days
Sperry, 1 crystal receiver with klystron oscillator in 60 days
7 Cathode-Ray Tube Display—RCA, 2 to 4 in 30 days

30 October 1940—MIT, Cambridge

Present: Alfred Loomis (Chairman)
Karl Compton, NDRC
Professor Ed Bowles, MIT
Ralph Bown, Bell Laboratories
Ralph Beal, RCA
Hugh Willis, Sperry
George Metcalf, GE
J A Hutcheson, Westinghouse
Melville Eastham, General Radio Corporation
Lee DuBridge, Radiation Laboratory
Ken Bainbridge, Radiation Laboratory
Majors Smith and Reeves, US Signal Corps

Notes

Examination of laboratory space available at MIT—10000 square feet within Main Building together with a 160 × 130 foot heated hangar at Boston Airport, with machine shop attached.
Review of contract arrangements as at previous meeting.
Discussion of scientific staff likely to be available.

7 November 1940—21 East 79th Street, New York City

Present: Alfred Loomis (Chairman)
Karl Compton, NDRC
Professor Ed Bowles, MIT
Ralph Bown, Bell Laboratories
Ralph Beal, RCA
Hugh Willis, Sperry
George Metcalf, GE
J A Hutcheson, Westinghouse
Melville Eastham, General Radio Corporation

Notes

Full day discussion of technical details of the first three Radiation Laboratory projects:

Project 1	10 centimetre airborne radar
Project 2	10 centimetre gun-laying radar
Project 3	Long-range navigation

15 November 1940—Carnegie Institution, Washington DC

Review of progress, but no details given in my diary.

Notes on subsequent meetings held in 1941 will be found in *Reports of the Tizard Mission,* Public Records Office, AVIA 10/1 to 4.

List of Abbreviations

Glossary of Radar Terms

AI	Air-Interception Radar. A radar installation in night fighters for detection and interception of enemy aircraft.
ASV	Air to Surface Vessel. An airborne radar for detection of ships and surfaced submarines.
CH	Home Chain. The chain of air-warning radars along the east and south coasts of Britain to provide early warning of the approach of enemy aircraft.
CHL	Home Chain Low. A network added later to give warning of low-flying aircraft.
CD	Coast Defence Radar.
DME	Distance Measuring Equipment. A radar system which gave a direct measurement of the distance of an aircraft from an airport.
GCI	Ground Control Interception Radar. A ground-based radar for the control of fighters equipped with AI.
GEE	A medium-range hyperbolic radio navigation system used by Bomber Command.
GL	Gun-Laying Radar.
IFF	A radar device for Identification of Friend from Foe.
LAB	Low Altitude Bombing attachment to an ASV radar.
LORAN	Long-Range Navigational system.
LRN	Long-Range Navigation.
LW/AW	Lightweight Air-Warning Radar.
OBOE	A highly accurate blind bombing aid used by RAF Pathfinder Squadrons.
PPI	Plan Position Indicator.
RDF	The first term used in Britain to describe a radar set.
SCR	Signal Corps Radio or Radar. Prefix to a series of numbers used by the US Signal Corps to designate radio and radar production equipment, e.g. SCR 584.
SLC	Searchlight Control Radar.
TRF	Tuned Radio-Frequency amplifier.

Offices and Establishments

ADRDE (Army) Air Defence Research and Development Establishment.
AMRE Air Ministry Research Establishment, Dundee and Swanage.
BRS Bawdsey Research Station.
CSIR Council for Scientific and Industrial Research, Australia.
CVD (Committee for) Coordination of Valve Development.
DCD Director of Communications Development.
DTM Department of Terrestial Magnetism, Washington.
FIU Fighter Interception Unit.
MAP Ministry of Aircraft Production, London.
MIT Massachusetts Institute of Technology, Cambridge, Massachusetts.
NDRC National Defence Research Council, USA.
NPL National Physical Laboratory, Teddington.
NRC National Research Council, Ottawa.
NRL Naval Research Laboratory, Washington, DC.
OSRD Office of Scientific Research and Development, USA.
TRE Telecommunications Research Establishment, Swanage and Malvern.

Bibliography

Radar Histories

Hartcup G 1970 *Challenge of War* (Newton Abbot: David and Charles)
Rowe A P 1948 *One Story of Radar* (Cambridge: Cambridge University Press)
Swords S S 1986 *Beginnings of Radar* (London: Peregrinus)
Watson Watt R A 1957 *Three Steps to Victory* (London: Odhams)
Watson Watt R A, Herd J F and Bainbridge-Bell L H 1933 *The Cathode Ray Tube in Radio Research* (London: HMSO)

Technical Manuals

Bell Laboratories Staff 1949 *Radar Systems and Components* ed. M J Kelly (New York: Van Nostrand)
Bowen E G ed. 1954 *A Textbook of Radar* 2nd edn (Cambridge: Cambridge University Press)
MIT Radiation Laboratory Series 1947–1953 ed. L N Ridenour **1–28** (New York: McGraw-Hill)
Proceedings of the Radiolocation Convention 1946 *Journal of Institution of Electrical Engineers* **93** (IIIA) No 1–4, 5–10 (London: IEE)

Other References

Allison D 1981 *New Eye for the Navy* (Washington DC: US Navy)
Guerlac H 1948 *Radar in World War II* 20 volumes (available from the Public Records Office, London)
Kinsey G 1981 *Martlesham Heath–Orfordness–Bawdsey* (Lavenham: Terence Dalton)
Sullivan W ed. 1984 *The Early Years of Radio Astronomy* (Cambridge: Cambridge University Press)

War Histories

Brandon L 1961 *Night Flyer* (London: William Kimber)
Busch H 1955 *U-Boats at War* (London: Putnam)
Chisholm R 1953 *Cover of Darkness* (London: Chatto and Windus)
Gunston W T 1976 *Night Fighters* (Cambridge: Patrick Squires)
Price A 1973 *Aircraft versus Submarine* (London: William Kimber)
Rawnsley J and Wright R 1957 *Night Fighter* (London: Collins)
Roskill S W 1954 *The War at Sea* 5 Volumes (London: HMSO)

Tizard Mission

No comprehensive account of the Tizard Mission has ever been written but references to the Mission are made in the following:

Baxter P J 1946 *Scientists Against Time* (Boston: Little, Brown)
Clarke R W 1965 *Sir Henry Tizard* (London: Methuen)
Guerlac H 1948 *Radar in World War II* 20 volumes (available from the Public Records Office, London)
Hartcup G and Allibone T E 1984 *Cockcroft and the Atom* (Bristol: Adam Hilger)

Index